Ancient and Archaic Genomes

Ancient and Archaic Genomes

Editors

David Caramelli
Martina Lari
Stefania Vai

MDPI • Basel • Beijing • Wuhan • Barcelona • Belgrade • Manchester • Tokyo • Cluj • Tianjin

Editors
David Caramelli
Department of Biology
University of Florence
Florence
Italy

Martina Lari
Department of Biology
University of Florence
Florence
Italy

Stefania Vai
Department of Biology
University of Florence
Florence
Italy

Editorial Office
MDPI
St. Alban-Anlage 66
4052 Basel, Switzerland

This is a reprint of articles from the Special Issue published online in the open access journal *Genes* (ISSN 2073-4425) (available at: www.mdpi.com/journal/genes/special_issues/ancient_genomes).

For citation purposes, cite each article independently as indicated on the article page online and as indicated below:

LastName, A.A.; LastName, B.B.; LastName, C.C. Article Title. *Journal Name* **Year**, *Volume Number*, Page Range.

ISBN 978-3-0365-2547-1 (Hbk)
ISBN 978-3-0365-2546-4 (PDF)

Contents

Editorial

Ancient and Archaic Genomes

Stefania Vai *, Martina Lari and David Caramelli

Department of Biology, University of Florence, 50122 Florence, Italy; martina.lari@unifi.it (M.L.); david.caramelli@unifi.it (D.C.)
* Correspondence: stefania.vai@unifi.it

The first data obtained from ancient DNA samples were published more than thirty years ago. During this time, methodological innovations signed by polymerase chain reaction (PCR) first, and next-generation sequencing (NGS) later allowed researchers to understand the molecular features of ancient DNA and to reconstruct even whole genomes from organisms that lived in the past. Dedicated facilities, specific experimental procedures, and bioinformatics pipelines are required to obtain reliable ancient DNA data taking into account preservation, degradation, and contamination issues [1]. Thanks to the already consolidated knowledge of DNA degradation processes and to the high-throughput sequencing methodologies, thousands of archaeological specimens (mostly bones and teeth, but also hairs, mummified soft tissues, coprolites, and vegetable materials) have been analyzed thus far at a genomic level, providing information for a deep understanding of evolutionary processes, anthropological questions, and archaeological reconstructions. Indeed, ancient DNA studies, providing direct data from the past, offer us the possibility to observe genetic variation through time, allowing us to go beyond the limits of inferential approaches relying on modern data only.

This Special Issue titled "Ancient and Archaic Genomes" collects original research articles that present different methods and aspects of the paleogenetic research applied to anthropological, archaeological, and historic questions. Interestingly, specific regional contexts and cultural aspects previously poorly studied from a genetic point of view are here investigated.

Through an experimental strategy based on PCR and Sanger sequencing, Gînguță and colleagues [2] investigated the maternal genetic diversity of medieval individuals from Transylvania (Romania). The mitochondrial DNA control region of 13 individuals from the Feldioara necropolis (12–13th century) was analyzed and compared with historical and modern populations. A high genetic variability was found, with all the individuals characterized by a different mitochondrial lineage, mostly carrying West Eurasian haplogroups and with a possible contribution related to the arrival of Hungarian conquerors at the end of the 9th century.

Thanks to NGS methodology, even complete mitochondrial genomes can be reconstructed, increasing the informative power of this uniparental marker. In this Issue, two contributions based on the analysis of complete or almost complete mitogenomes are dedicated to ancient human samples, while one article is focused on animal specimens.

Kusliy et al. [3], indeed, present a study on the Mongolian horse, one of the most ancient horse breeds. They obtained nearly complete mitochondrial genomes from six samples of the Khereksur and Deer Stone cultures (late 2nd to first third of the 1st millennium BC) and from the Xiongnu culture (1st century BC to 1st century AD). A phylogenetic analysis revealed genetic continuity between the Mongolian horse populations of the three ancient cultures, and a comparison with other ancient, historical, and modern mitogenomes of horses indicated close relationships with indigenous breeds of the Middle East, Eastern and Central Asia, and the Mediterranean region.

In the context of the study of past human population dynamics, Fontani and colleagues [4] present the first ancient molecular data from the Calabria region (southwestern Italy):

Citation: Vai, S.; Lari, M.; Caramelli, D. Ancient and Archaic Genomes. *Genes* **2021**, *12*, 1411. https://doi.org/10.3390/genes12091411

Received: 30 August 2021
Accepted: 8 September 2021
Published: 13 September 2021

Publisher's Note: MDPI stays neutral with regard to jurisdictional claims in published maps and institutional affiliations.

an important contribution to fill the gap in the limited knowledge of prehistoric genetic variability in the Italian Peninsula. Two mitochondrial genomes were obtained from a Middle Bronze Age mass grave in a karstic cave called Grotta della Monaca. Possible affinities with other Italian and European ancient populations are highlighted through a phylogenetic analysis, providing a starting point for the study of population dynamics and migrations in Southern Italy.

Another research article presenting data on whole mitochondria is focused on 10–11th century remains in the Carpathian Basin. Maár et al. [5] analyzed more than 200 new mitogenomes for the commoner population in order to compare them with the data available for the immigrant elite including conquering Hungarians. Phylogenetic analysis and haplogroup- and sequence-based methods provide a first description of this population and the relationships with other ancient Eurasian groups, highlighting differences and possible admixture with the eastern immigrants.

A more accurate picture of the population dynamics in these European regions, firstly described through mitochondrial genetic variability, might be provided by future genome-wide studies. In fact, mitochondrial data alone provide a partial view of the genetic history of a population. A greater contribution than NGS technology provided to ancient DNA studies is represented by the possibility to obtain reliable autosomal data that can reveal more details not only at a population but also at an individual level.

Autosomal data are often obtained in ancient degraded samples through a target enrichment approach focused on genotyping a set of informative SNPs. Based on this approach, mitochondrial and nuclear SNP data are presented in a case study showing how paleogenomic analysis of ancient human remains can provide information useful for individual characterization and archaeological reconstruction [6]. The minimum number of individuals, their sex and phenotypic traits, and the kin relationships between them were estimated for samples found in a complex scenario from the Neolithic time in Poland: a multiple secondary burial where skeletal remains of several individuals were intentionally fragmented and mixed. The analysis of molecular damage also highlighted the presence of modern remains added in historical time to the burial boundary, helping to clarify the reconstruction of the exploitation of this site through time.

The SNP selection strategy is often the only way to obtain nuclear genome data for highly degraded samples with a low content of endogenous DNA. However, when allowed by DNA preservation, a whole-genome sequencing (WGS) strategy allows performing sophisticated analysis that helps to understand past demographic scenarios and infer details about populations' structure and interactions.

Vizzari et al. [7] tested the two main out-of-Africa hypotheses through an approximate Bayesian computation approach based on the random forest algorithm. It is still under debate, indeed, whether anatomically modern humans (AMH) left Africa through a single dispersion event or in two main waves (first toward southern Asia and Australo-Melanesia, and later through a northern route). By comparing simulated data with real genomic variation observed by analyzing high-coverage genomes of archaic (Denisovan and Neandertal) and modern populations, they show that a model of multiple dispersals is four-fold as likely as the alternative single-dispersal model. Modern Australo-Melanesians derive from a migration from Africa that may have occurred around 74,000 years ago, while Eurasians derive from a second dispersal event dating back to around 46,000 years ago.

We believe that this Special Issue, presenting different methodological approaches and applications, will be a useful resource for both students and young researchers who are interested in ancient DNA studies.

Author Contributions: S.V., M.L. and D.C. designed and wrote the manuscript. All authors have read and agreed to the published version of the manuscript.

Funding: This research received no external funding.

Conflicts of Interest: The authors declare no conflict of interest.

References

1. Orlando, L.; Allaby, R.; Skoglund, P.; Der Sarkissian, C.; Stockhammer, P.W.; Ávila-Arcos, M.C.; Fu, Q.; Krause, J.; Willerslev, E.; Stone, A.C.; et al. Ancient DNA analysis. *Nat. Rev. Methods Primers* **2021**, *1*, 14. [CrossRef]
2. Gînguță, A.; Rusu, I.; Mircea, C.; Ioniță, A.; Banciu, H.L.; Kelemen, B. Mitochondrial DNA Profiles of Individuals from a 12th Century Necropolis in Feldioara (Transylvania). *Genes* **2021**, *12*, 436. [CrossRef] [PubMed]
3. Kusliy, M.A.; Vorobieva, N.V.; Tishkin, A.A.; Makunin, A.I.; Druzhkova, A.S.; Trifonov, V.A.; Iderkhangai, T.-O.; Graphodatsky, A.S. Traces of Late Bronze and Early Iron Age Mongolian Horse Mitochondrial Lineages in Modern Populations. *Genes* **2021**, *12*, 412. [CrossRef] [PubMed]
4. Fontani, F.; Cilli, E.; Arena, F.; Sarno, S.; Modi, A.; De Fanti, S.; Andrews, A.J.; Latorre, A.; Abondio, P.; Larocca, F.; et al. First Bronze Age Human Mitogenomes from Calabria (Grotta Della Monaca, Southern Italy). *Genes* **2021**, *12*, 636. [CrossRef] [PubMed]
5. Maár, K.; Varga, G.I.B.; Kovács, B.; Schütz, O.; Maróti, Z.; Kalmár, T.; Nyerki, E.; Nagy, I.; Latinovics, D.; Tihanyi, B.; et al. Maternal Lineages from 10–11th Century Commoner Cemeteries of the Carpathian Basin. *Genes* **2021**, *12*, 460. [CrossRef] [PubMed]
6. Vai, S.; Diroma, M.A.; Cannariato, C.; Budnik, A.; Lari, M.; Caramelli, D.; Pilli, E. How a Paleogenomic Approach Can Provide Details on Bioarchaeological Reconstruction: A Case Study from the Globular Amphorae Culture. *Genes* **2021**, *12*, 910. [CrossRef] [PubMed]
7. Vizzari, M.T.; Benazzo, A.; Barbujani, G.; Ghirotto, S. A Revised Model of Anatomically Modern Human Expansions Out of Africa through a Machine Learning Approximate Bayesian Computation Approach. *Genes* **2020**, *11*, 1510. [CrossRef] [PubMed]

Article

Traces of Late Bronze and Early Iron Age Mongolian Horse Mitochondrial Lineages in Modern Populations

Mariya A. Kusliy [1,*], Nadezhda V. Vorobieva [1], Alexey A. Tishkin [2], Alexey I. Makunin [1], Anna S. Druzhkova [1], Vladimir A. Trifonov [1], Tumur-O. Iderkhangai [3] and Alexander S. Graphodatsky [1]

1 Department of the Diversity and Evolution of Genomes, Institute of Molecular and Cellular Biology SB RAS, 630090 Novosibirsk, Russia; vorn@mcb.nsc.ru (N.V.V.); alex.makunin@gmail.com (A.I.M.); rada@mcb.nsc.ru (A.S.D.); vlad@mcb.nsc.ru (V.A.T.); graf@mcb.nsc.ru (A.S.G.)

2 Department of Archaeology, Ethnography and Museology, Altai State University, 656049 Barnaul, Russia; tishkin210@mail.ru

3 Department of Archaeology, Ulaanbaatar State University, Ulaanbaatar 13343, Mongolia; iderkhangai83@gmail.com

* Correspondence: kusliy.maria@mcb.nsc.ru

Abstract: The Mongolian horse is one of the most ancient and relatively unmanaged horse breeds. The population history of the Mongolian horse remains poorly understood due to a lack of information on ancient and modern DNA. Here, we report nearly complete mitochondrial genome data obtained from five ancient Mongolian horse samples of the Khereksur and Deer Stone culture (late 2nd to 1st third of the 1st millennium BC) and one ancient horse specimen from the Xiongnu culture (1st century BC to 1st century AD) using target enrichment and high-throughput sequencing methods. Phylogenetic analysis involving ancient, historical, and modern mitogenomes of horses from Mongolia and other regions showed the presence of three mitochondrial haplogroups in the ancient Mongolian horse populations studied here and similar haplotype composition of ancient and modern horse populations of Mongolia. Our results revealed genetic continuity between the Mongolian horse populations of the Khereksur and Deer Stone culture and those of the Xiongnu culture owing to the presence of related mitotypes. Besides, we report close phylogenetic relationships between haplotypes of the Khereksur and Deer Stone horses and the horses of indigenous breeds of the Middle East (Caspian and Iranian), China (Naqu, Yunnan, and Jinjiang), and Italy (Giara) as well as genetic similarity between the Xiongnu Mongolian horses and those of the most ancient breeds of the Middle East (Arabian) and Central Asia (Akhal-Teke). Despite all the migrations of the Mongolian peoples over the past 3000 years, mitochondrial haplogroup composition of Mongolian horse populations remains almost unchanged.

Keywords: ancient DNA; mitochondrial DNA; Mongolian horse; phylogeography

Citation: Kusliy, M.A.; Vorobieva, N.V.; Tishkin, A.A.; Makunin, A.I.; Druzhkova, A.S.; Trifonov, V.A.; Iderkhangai, T.-O; Graphodatsky, A.S. Traces of Late Bronze and Early Iron Age Mongolian Horse Mitochondrial Lineages in Modern Populations. *Genes* **2021**, *12*, 412. https://doi.org/10.3390/genes12030412

Academic Editor: David Caramelli

Received: 31 January 2021
Accepted: 9 March 2021
Published: 12 March 2021

1. Introduction

The Mongolian horse breed is referred to as a "landrace breed" because these horses are free-ranging and to a lesser extent have experienced selective pressures applied by breeders [1]. Mongolian horse populations are believed to have been the source for many breeds across Asia [2], including Tuva [1], Yunnan [3], and Jeju [4–6]. The assumption that the ancient Mongolian horse population might have been ancestral to many modern horse breeds [3] is also supported by the highest levels of within-breed diversity according to genome-wide autosomal, microsatellite, protein, and mitochondrial polymorphism data [1,3,7,8]. Even though the Mongolian horse is one of the most ancient breeds and is considered ancestral for many other breeds, very little research on its ancient DNA has been carried out, though abundant osteological material has been collected [9–11]. Only two mitogenomes of historical Mongolian horses are available in GenBank, and even these mitogenomes are not older than the 20th century. A few ancient horse populations of

Mongolia have been investigated at the mitochondrial level [4,12,13]. These studies have determined hypervariable region mitotypes of three ancient horses found in burials of the Early Iron Age Pazyryk archaeological sites (Olon-Kurin-Gol-6 and Olon-Kurin-Gol-10) in Northwestern Mongolia [13]. According to the classification of Cieslak et al. [12], the three horses carry the haplotypes that were widespread in the Iron Age in China, Southwest Siberia, and Kazakhstan [13]. A comparison of mitochondrial control region diversity between ancient and modern Mongolian horse populations has revealed that the Pazyryk horse haplotypes also occur in modern populations. However, modern populations also contain other haplotypes. It has been shown that many East Asian mitochondrial lineages of Bronze and Iron Age were missing in other regions of the Eurasian steppe (West Siberia, South Siberia, Kazakhstan), possibly owing to the isolation of Mongolian and Chinese wild horses by the Altai Mountains and the Takla Makan and Gobi deserts [12]. To date, only one study has been conducted in which complete mitochondrial genomes were obtained for the Mongolian ancient horses of the Khereksur and Deer Stone culture and the Xiongnu culture. Phylogeographic reconstructions in this research showed that Mongolian horses of the Bronze Age Khereksur and Deer Stone culture were related to the Bronze Age horses of Moldova and Germany and 5500-year-old Botai culture horses of the Central Asian steppes. The results of that study also revealed that the horses of the Iron Age Xiongnu culture were closely related to the horses of the Neolithic period from the Orenburg region of Russia and those of the Gallo-Roman period of classical antiquity from France [4].

We carried out a molecular genetic study of bone samples of ancient horses associated with the Ganga Tsagaan ereg (late 12th to mid-10th century BC) and the Ereen hailaas (1st century BC to 1st century AD) archaeological sites, belonging to the Khereksur and Deer Stone archaeological culture and the Xiongnu archaeological culture, respectively. The above-mentioned archaeological sites are located close to each other in the valley of the Egiin gol river in the territory of Northern Mongolia. The Ganga Tsagaan ereg archaeological site belongs to the Khereksur and Deer Stone culture, which represents the most archaic nomadic empire [14]. No settlement of this culture has been found yet [14,15]. Most of the investigated kurgans (khereksurs) are characterized by the presence of so-called altars (small stone mounds), under which a horse skull as well as some cervical vertebrae bones and hooves have been found. This ritual (symbolic) action implied an accompanying burial of a horse for the afterlife. No deer stones have been found at the Ganga Tsagaan ereg site, thus indirectly indicating an earlier time point of the khereksur construction in comparison with those with deer stones [15]. Archaeological and paleozoological data indicate that clear evidence of widespread use of domesticated horses in Mongolia dates back to the end of the 2nd millennium BC and is associated with the Khereksur and Deer Stone culture [16,17]; ancient people of this culture mainly bred horses, sheep, and cattle [16]. The Ereen hailaas archaeological site (Xiongnu culture) is related to the period when the location of the Shanyu (leader of the Xiongnu Empire) was moved from the Chinese Han Empire far to the north, and the Huns began to explore the territory of modern Mongolia as well as Buryatia (Russia) and Transbaikalia (Russia). The Xiongnu society was characterized by an economy based on nomadic animal husbandry. The nomads of the Xiongnu Empire bred animals traditional for the Eurasian steppes: horses, sheep, goats, camels, and cattle. Of all these species, horses had the highest economic and military significance and had a special place in the culture of ancient nomads [18]. The results of paleofaunistic studies have shown that by their exterior features, most of the Xiongnu horses were similar to the horses of the Mongolian type [19]. However, it cannot be ruled out that some of the horses of the elite were Central Asian horses (Akhal-Teke horse) because similar horses are depicted on burial mound drapery from the Noin-Ula site (Xiongnu culture) [20].

The aim of this study was to determine mitochondrial diversity of ancient Mongolian horse populations of the Khereksur and Deer Stone (Ganga Tsagaan ereg site) archaeological culture and the Xiongnu (Ereen hailaas site) archaeological culture, their continuity, and phylogeographic relationships by means of mitogenome target-enriched high-throughput sequencing data. As a result, we report nearly complete mitogenomes from six ancient Mongolian horses, phylogeographic analysis of which, together with previously published horse mitogenomes, revealed the presence of similar mitotypes in the studied ancient Mongolian populations and close phylogenetic relationships between the horses studied here and modern horses of the most ancient and indigenous breeds of Central and Eastern Asia, the Middle East, and the Mediterranean region.

2. Materials and Methods

2.1. Information about the Samples

Five osteological specimens of ancient horses from Mongolian archaeological site Ganga Tsagaan ereg (Khereksur and Deer Stone culture, late 12th to mid-10th century BC) and one tooth specimen of an ancient horse from Mongolian archaeological site Ereen hailaas (Xiongnu culture, 1st century BC to 1st century AD) were investigated in this work (detailed information about the samples is given in Table S1).

2.2. Ancient-Mitogenome Sequencing

All experiments were conducted at the Institute of Molecular and Cellular Biology ofthe Siberian Branch of the Russian Academy of Sciences (SB RAS) (Novosibirsk, Russia), in a special laboratory focused only on the research on ancient DNA, with maximum protection against contamination (use of special protective clothing, surface treatment with DNAZap nucleic-acid–degrading solutions, and ultraviolet irradiation of laboratory premises). All ancient bone specimens were subjected to UV irradiation (30 min on each side of a bone); 1 cm^2 of bone (0.3 g) was sawed off with a diamond disk of an electric drill, the surface layer was removed, and the bone was crushed into a powder in a metal mortar. Ancient-DNA extraction was performed following the protocol described in the article of Yang et al. [21] with one modification made by Sanderson, Radley, Mayton [22]: lysis buffer contained NH_4-EDTA for reducing decalcification time from 18–24 to 1.5–3.0 h at a standard concentration of 300 mg of the bone powder per 5 mL of the buffer at 55 °C. Genomic libraries of ancient DNA fragments were obtained using the TruSeq Nano DNA Sample Preparation Kit (Illumina, San Diego, CA, USA) according to the manufacturer's protocol: High Sample Protocol with a few modifications: samples were purified with the MinElute PCR Purification Kit (Qiagen, Hilden, Germany) instead of purification on the AMPure XP beads (Illumina), and 12 cycles of library amplification were performed.

Two rounds of library enrichment were carried out by hybridization with biotinylated modern *Equus caballus* mitochondrial DNA (mtDNA) immobilized on DynabeadsVR Streptavidin magnetic particles (Life Technologies), according to a previously published method [23] with modifications, described in the article of Vorobieva et al. [24]. Amplification of the enriched libraries was carried out in a volume of 50 μL containing 1× Phusion HF Buffer, 0.2 mM each dNTP, 1 μM each primer for library fragment adapters (SuD Nano DNA Library Prep Kit for Illumina), 1 unit of Phusion DNA polymerase. Cycling conditions included 30 s initial denaturation at 95 °C; followed by 20 cycles of 20 s denaturation at 98 °C, 20 s primer annealing at 65 °C, and 20 s elongation at 72 °C; with final 5 min elongation at 72 °C.

The obtained libraries were quantified by real-time PCR in the presence of dye SYBR Green 1 and reference dye for quantitative PCR (ROX), using a PCR kit: 2.5× reaction mix (Syntol), according to the manufacturer's protocol. Contamination was monitored through extraction, library, and amplification blanks, which all yielded negative results. Paired-end sequencing of the enriched libraries was performed on the MiSeq platform (Illumina) with the MiSeq v2 Reagent Kit (300 cycles, 2 × 150 bp).

2.3. Sequence Data Analysis

For the initial sequencing data processing, PALEOMIX BAM Pipeline v1.3.2 was used [25]. Reads were trimmed and collapsed (AdapterRemoval v.2.2.2) [26]. Sequencing read alignments against reference sequences of the horse mitogenome (GenBank accession No.: NC_001640.1) and the human mitogenome for contamination control (GenBank accession No.: NC_012920.1) were performed using sequence aligner bwa v0.7.15 [27] with the following parameters: algorithm, mem; minimum mapping quality, 25; and the FilterUnmappedReads option. The alignment improvement included filtering of PCR duplicates (paleomix rmdup_collapsed). To obtain more reliable sequences of the mitochondrial genomes, we performed rescaling of the base quality scores in the BAM alignment file, according to the base's probability of being affected by postmortem damage, and constructed a postmortem DNA damage model using the MapDamage v2.2.0 computational framework [28]. To remove human DNA contamination, a custom script (https://github.com/lca-imcb/lca-ngs/blob/master/contam_filter.py (accessed on 19 February 2021)) based on mapping quality comparison was employed. Consensus sequences were called using a minimum depth of coverage of 2 and filtering for base quality ≥30; all variants alternative to the reference sequence were confirmed on bioinformatics software platform Geneious Prime v2020.2.4 (https://www.geneious.com (accessed on 19 February 2021)) with a set threshold of the highest quality (bases had to match at least 60% of total adjusted quality). The depth and width of the mitochondrial genome coverage were determined on software platform Geneious Prime v2020.2.4.

2.4. Phylogenetic Analysis

Multiple alignment of the mitogenome consensus sequences was conducted using multiple sequence alignment program MAFFT v7.407 [29]. To select best-fit partitioning schemes and models of molecular evolution for phylogenetic analyses, we utilized PartitionFinder v2.1.1 [30]. In the horse mitochondrial genome alignment, we identified 5 partitions with the following evolutionary models of nucleotide substitutions: HKY + I for the second codons of protein-coding genes; HKY + G + I for RNA-coding genes, first and third codons of protein-coding genes; and GTR + G + I for hypervariable regions. A time-measured Bayesian phylogenetic tree was constructed including 206 horse mitogenomes via the above-mentioned partitioning schemes and evolutionary models on an advanced software platform for Bayesian evolutionary analysis, BEAST v1.10.4 [31] (50 million generations of the Markov Chain, sampling frequency: 1000, the first 10% of the trees were discarded as burn-in), clock model: strict clock [32], tree prior: coalescent (constant population size) [33,34], and the tree model: random starting tree. The branch divergence times were derived using the ages of all ancient samples as internal calibrations at the tips of the evolutionary tree. All specimens used for the phylogenetic reconstructions were dated either in calibrated years before the present or years before the present associated with the corresponding archaeological context (Table S2). The BEAST configuration file was generated in BEAUti v1.10.4, and the BEAST output was analyzed in Tracer v1.7.1 [35]. In the latter, effective sample sizes for all the traces were greater than 800, and posterior probability density was bell-shaped, suggesting that the parameter estimates reached convergence. To summarize the information contained within our sampled trees, we used a tool called TreeAnnotator v1.10.4 [36]. The annotated tree was visualized using the FigTree v1.4.4 software (http://tree.bio.ed.ac.uk/software/figtree/ (accessed on 22 February 2021)).

2.5. Data Availability

All studied bone and tooth samples (Er1, Gan1, Gan3, Gan11, Gan14, and Gan18) of ancient horses are stored in a repository located in the Department of Archaeology of Ulaanbaatar State University, Ulaanbaatar, Mongolia. The samples were provided by a coauthor of this article, Dr. Tumur-Ochir Iderkhangai. The present study is compliant with all relevant laws and regulations. All the experiments were approved by the Ethics Committee on Animal and Human Research at the Institute of Molecular and Cellular Biology, Russia (permit No. 01/20 of 11 February 2020).

The next-generation sequencing data were submitted to the Sequence Read Archive database (SRA) and are available under BioProject accession number PRJNA694234 (https://www.ncbi.nlm.nih.gov/bioproject/?term=PRJNA694234 (accessed on 23 February 2021)). GenBank accession numbers of the mitogenome consensus sequences are as follows: MW534078 for sample Er1, MW534079 for sample Gan1, MW534080 for sample Gan3, MW534081 for sample Gan11, MW534082 for sample Gan14, and MW534083 for sample Gan18.

3. Results

3.1. Authenticity of Ancient DNA Data

As already determined, the main characteristics of ancient DNA molecules are cytosine deamination resulting in an increased C → T substitution rate and an increased complementary G → A substitution rate towards 5″ and 3″ read termini, respectively, as well as DNA degradation to small fragments, generally below 100 bp [37]. Using the MapDamage v2.2.0 computational framework, postmortem damage and fragmentation patterns in the obtained ancient DNA libraries were evaluated (Figures S1 and S2, respectively). All the ancient samples studied here showed expected nucleotide misincorporation signatures of postmortem DNA damage, proving the authenticity of the ancient DNA. However, the 5″ C-to-T damage profile was less pronounced due to peculiarities of the library preparation procedure (related to the blunt-end repair process, when the polymerase removes the 3″ overhangs and fills in the 5″ overhangs of the fragments). The average size of DNA fragments of the studied libraries varied from 60 to 90 bp, which also points to the antiquity of the target DNA. Among the studied samples, the youngest sample of the Xiongnu culture (Er1) has the largest average length of DNA fragments, indicating that in our sample set, the degree of DNA degradation depends on sample age. However, the conditions of sample preservation were slightly different. The bone remains of the Er1 horse were located in a pit along with the osteological material of other animals, while the bone specimens of each Ganga Tsagaan ereg horse were buried in a separate altar and were well isolated from other structures of the ritual complex [15]. We believe that the lower genome coverage of the Er1 sample results from a higher level of exogenous DNA that can be attributed to the difference in sample location (the pit rather than an altar for other samples). The amount of human contamination removed using the custom script (https://github.com/lca-imcb/lca-ngs/blob/master/contam_filter.py (accessed on 23 February 2021)) ranged from 0 to 150 reads per sequencing library.

3.2. Characteristics of the Ancient Mongolian Horse Mitochondrial Genomes

The width of coverage of the obtained mitochondrial genomes of the ancient horses varies from 97.3% to 99.8% of the reference sequence length, and the average depth of coverage is between 6.7-fold and 85.9-fold. The ratio of the unique mapped collapsed reads to the total number of collapsed reads is 1.2% on average. We used the obtained sequences of the nearly complete mitochondrial genomes of the ancient Mongolian horses in our phylogenetic reconstructions. Detailed main characteristics of the obtained mitogenomes are shown in Table 1. Coverage depth of the obtained mitogenomes of ancient Mongolian horses is visualized in Figure S3.

Table 1. The main characteristics of the mitochondrial genomes of the ancient horses of Mongolia.

Sample Name	Number of Collapsed Reads	Number of Unique Mapped Collapsed Reads	Mitogenome Width of Coverage, %	Mitogenome Average Depth of Coverage	Ancient Library Average Fragment Size	Terminal Library Fragment Deamination, %
Er1	188,493	917	97.3	6.7	90	14.49
Gan1	714,656	3143	99.1	18.6	60	24.31
Gan3	743,175	6919	99.8	41.8	63	25.88
Gan11	549,992	13,893	99.7	85.9	68	28.51
Gan14	511,882	11,643	99.8	66.7	65	33.77
Gan18	582,595	3322	99.5	18.5	59	25.26

3.3. Phylogenetic Reconstructions

By the Bayesian method, a phylogenetic tree was constructed (Figure 1) based on the obtained mitogenome sequences and the mitogenome sequences of modern horses of different breeds investigated in the study by Achilli et al. [38], in which there have been identified 18 major mitochondrial haplogroups (A–R) and 83 mitotypes. Due to a lack of mitogenome sequences from ancient and modern horses of Asian breeds, this set of sequences was supplemented with mitogenomes from the GenBank sequence database, with special attention paid to the horse breeds that are considered closely related to the Mongolian horse (Tuva, Yakutian, Yunnan, and Jeju [1,3,4,6]) and those with origins in East Asia. Information on the mitogenome sequences used to build the phylogenetic tree is given in Table S2. Initial trees that display the 95% confidence intervals around the mean divergence time are presented in Figures S4 and S5 (with and without *Equus asinus* (outgroup), respectively).

Almost all ancient horses outside Mongolia shown in the phylogenetic tree must have been wild (*Equus ferus* lineage). They are located in the tree branches not belonging to any domesticated horse haplogroups identified by Achilli et al. [38] and therefore are most likely to be extinct clades of evolution. All these extinct lineages sampled so far originated from the territories of the Ural Mountains and Western and Eastern Siberia (Russia) [39,40]. According to the literature data, horses of the extinct *Equus lenensis* lineage inhabited the territories of Eastern and Western Siberia from the Late Pleistocene to the Middle Holocene [4,41]. The historical and modern Przewalski's horses (*Equus ferus przewalskii*) and their hybrids with domestic horses (*Equus ferus caballus*) are also not related to the ancient Mongolian horses analyzed here. None of the studied ancient horses of Mongolia are located in the same clade with Lena horses (*Equus lenensis*) in the phylogenetic tree. Consequently, the mitotypes of the ancient Mongolian horses are within mitochondrial diversity of domestic horses.

Ancient horses from the Ganga Tsagaan ereg archaeological site (Khereksur and Deer Stone culture) were found to belong to mitogroups G, L, and M identified by Achilli et al. [38] and the ancient horse from the Ereen hailaas site (Xiongnu culture) was shown to belong to mitogroup G. Consequently, this haplogroup was present in the ancient horse populations of Mongolia, both in the Khereksur and Deer Stone culture and in the Xiongnu culture, thus indicating certain continuity between the horse populations of these Mongolian archaeological cultures. Comparing the obtained results with those of a recently published study [24], we can conclude that the studied ancient Mongolian horse populations of the Late Bronze Age and the Early Iron Age were less diverse in mitochondrial composition than the Ukok Altai horses of the Early Iron Age. Six Ukok horses have been assigned to five different mitogroups: A, I, N, P, and Q, as classified by Achilli et al. [38].

According to the published data, haplogroup G is more common in Asia and has lower prevalence in the Middle East and Southern Europe. Haplogroup L is more common in Europe, becoming less and less common toward the East, albeit containing many mitotypes of different Chinese breeds. Haplogroup M has the same prevalence in Europe and Asia. Haplogroups G and M were more common among ancient horse populations as compared to modern ones [38]. Therefore, it is not surprising that the four ancient horses of Mongolia under study fell into these haplogroups.

We noticed a connection between the khereksur number and the mitogroup of horses whose bone remains were found in the altars, next to the khereksurs. Horses of the Ganga Tsagaan ereg site associated with the same khereksur belong to the same mitochondrial haplogroup. Horses Gan14 and Gan18 belonging to mitogroup L were buried in altars number 10 and number 14, respectively, next to khereksur 1-082. Horses Gan1 and Gan3 belong to the same haplogroup M, and their bone remains were found in altars number 2 and 4, respectively, next to khereksur 1-078 [15]. We inferred that most likely these studied ancient horses of different haplogroups belonged to different people and were bred independently of each other.

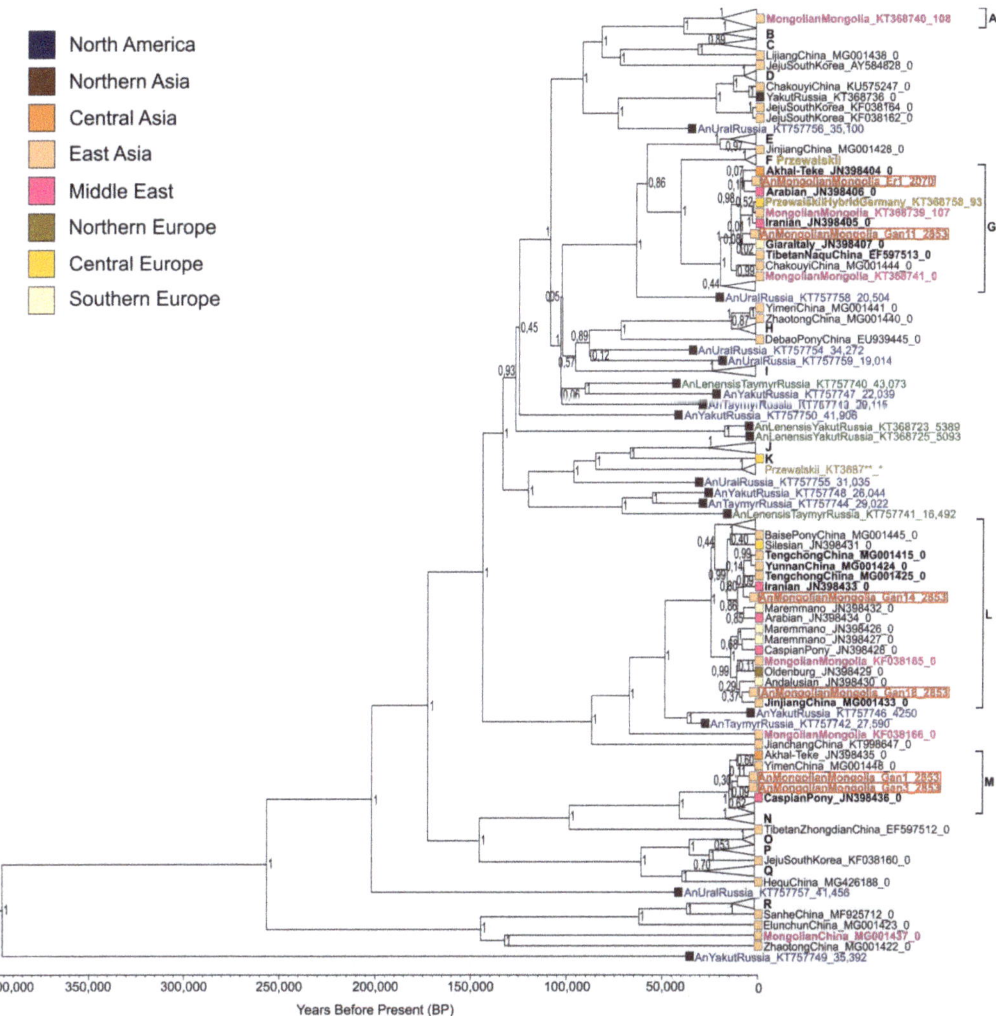

Figure 1. The Bayesian phylogenetic tree of the ancient Mongolian horses studied here and ancient and modern horses of different origins. The tree was constructed based on the alignment of 6 sequences of mitogenomes obtained here and 200 previously published ones (Table S2). Mitochondrial haplogroups A–R by classification of Achilli et al. [38] are marked in black. The red color of a sample name denotes the ancient horses of Mongolia studied here, the green color indicates those belonging to the *Equus lenensis* lineage, golden means belonging to Przewalski's horse, blue denotes ancient horses outside Mongolia, pink means modern and historical (20th century) horses of Mongolia, and black indicates modern horses outside Mongolia. Sample names consist of three parts separated by underscores. The first part is a species affiliation or horse breed name and geographic origin of the sample, the second part is a GenBank accession number, and the third part is sample age. The color palette of the squares indicates the geographical origin of samples. The tree was rooted by means of a published donkey (*Equus asinus*) mitochondrial genome (GenBank accession No.: NC_001788.1) (not displayed). Numbers near the tree branch nodes indicate posterior probability of the topology obtained by the Bayesian method. The time divergence scale at the bottom of the figure is a timeline where the dates are expressed in years before the present. Tree nodes with modern horses, not considered in the Discussion section, were collapsed for better visualization.

3.4. Time of Haplotype Divergence

On the basis of the constructed Bayesian phylogenetic tree (Figure 1), we estimated the time of divergence of the identified mitotypes and the closely related ones. Age estimates of the relevant nodes with 95% HPD (height posterior density) are shown in Table 2. Most Mongolian ancient horse mitotypes and the closest genetically related mitotypes diverged ~5000 years ago (median age estimate). Only the ancestral mitotype of two ancient Mongolian horses of the Khereksur and Deer Stone culture and the (closest to them) modern horse of the Caspian pony breed existed ~10,000 years ago (median age estimate). The extinct genetic lineages of wild horses from Eastern and Western Siberia located in the basal positions in relation to the main haplogroups and the lineages of domesticated horses diverged within the last 50–390 thousand years (median age estimates). Domesticated horses of haplogroup L and wild horses from Yakutia (the youngest sample of a wild horse: 4300 years [39]) and from the Taymyr Peninsula have the latest divergence time (median age estimate: 50,000 years). We can conclude that the divergence times of mitotypes of the studied Mongolian ancient horses and the closest related modern horse mitotypes belong the time span from the Upper Paleolithic to the Early Iron Age (95% HPD: [2330.9; 14,467.8]).

Table 2. BEAST age estimates of the relevant nodes with the ancient Mongolian horses studied in the phylogenetic tree constructed here.

Node Name	Median Age Estimate	95% HPD (Height Posterior Density)
Er1_Arabian_Akhal-Teke	3885.2	[2330.9; 6645.42]
Gan11_Giara_Naqu	5056.25	[3105.38; 7914.36]
Gan14_Iranian_Tengchong_Yunnan_Tengchong	5687.42	[3150.41; 9072.17]
Gan18_Jinjiang	4432.33	[2853.16; 7625.08]
Gan1_Gan3_Caspian	9416.83	[5911.57; 14,467.8]

4. Discussion

4.1. Phylogeographic Relationships

The geographic origin data on the horse mitotypes closely related to the ancient Mongolian ones obtained here are depicted in Figure 2.

After detailed consideration of the phylogenetic relationships within haplogroup G, it becomes clear that ancient Mongolian horse Er1 of the Xiongnu culture falls into the same clade with modern horses of the Arabian and the Akhal-Teke Middle Eastern breeds. The Arabian horse breed is one of the oldest in the world. A recent study involving equine single-nucleotide polymorphism (SNP) arrays and whole-genome resequencing uncovered multiple origins of Arabian horses from the Middle East region [42]. The Akhal-Teke horse breed is also one of the most ancient. This is a breed of oriental origin, indigenous to Central Asia in the area of Turkmenistan from the Caspian coast to the Fergana Valley. According to historical data, this breed was developed several thousand years ago [43]. Another study, by means of historical, faunal, genetic, and iconographic data, has shown that the breed ancestral to Akhal-Teke horses is the horses ridden by the Pazyryk chieftains (4th to 2nd century BC). It has been suggested that Akhal-Teke horses were obtained by crossing the domesticated horse from the Middle Volga with the tarpan (*Equus ferus ferus*) of the Eurasian steppes [44]. From this information, we can assume that the ancient Mongolian horse of the Xiongnu culture studied here is closely related to modern horses of the most ancient breeds with origins in the Middle East and Central Asia.

One of the ancient horses, Gan11 from the Ganga Tsagaan ereg archaeological site of the Khereksur and Deer Stone culture, is located in the same clade with modern horses of the Tibetan Naqu breed and the Italian Giara breed. The Naqu horse is a small indigenous breed of China, it includes Tibetan-type horses with strong adaptation to the harsh environment of the Tibet region and areas adjacent to the Qinghai–Tibet Plateau [45]. The genetic kinship between the Tibetan and the Mongolian horse breeds has also been proven in a study based on the analysis of microsatellite markers [46]. The Giara horse is one of the 15 extant native Italian breeds and is geographically isolated on the island of Sardinia [47]. Based on phylogenetic reconstruction, it can be concluded that the ancient Mongolian horse Gan11 is closely related to the modern horses of the native breeds of the Qinghai-Tibet Plateau in East Asia and the island of Sardinia in the Mediterranean region; this finding is quite interesting because these regions are geographically very distant.

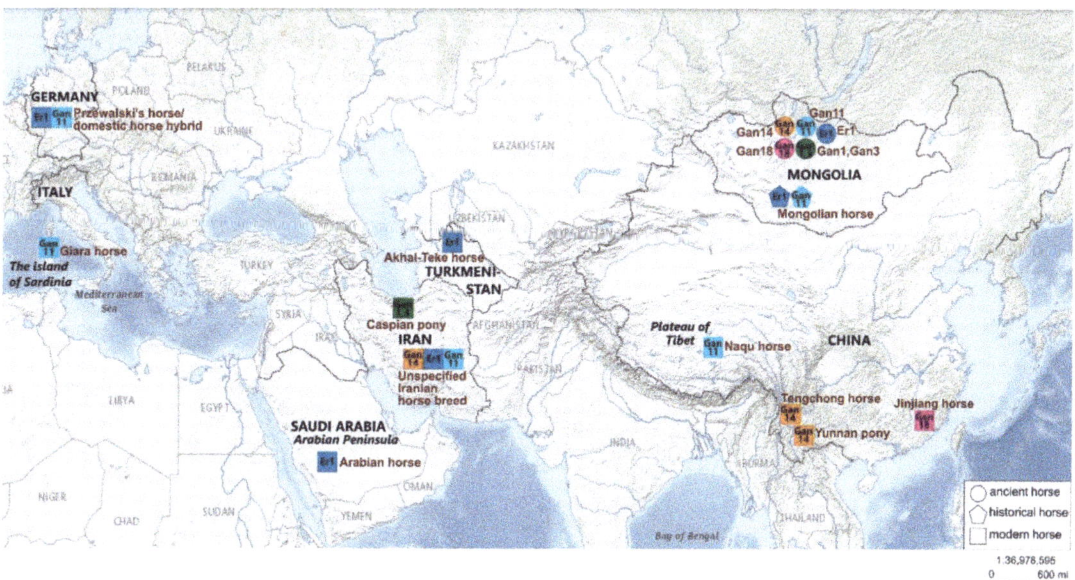

Figure 2. Geographical origin of the mitotypes closely related to the ancient ones obtained here. Circles represent ancient horses, and squares mean modern horses. Identical colors of circles and squares indicate similar mitotypes. Brown names are the names of ancient horse specimens and horse breeds. Black names and boundaries denote geographical objects within which the origin regions of closely related mitotypes are located. The figure was prepared based on the map in the USGS (United States Geological Survey) National Map Viewer (public domain) resource; it is similar but not identical to the original image and is therefore for illustrative purposes only.

Given that the clades with ancient horses Er-1 and Gan-11 are not well supported, we would like to consider their sister clade, which includes the modern Mongolian horse, a hybrid of Przewalski's horse, a modern horse, and a horse of an unspecified Iranian breed. The similarity of the mitotypes of modern and ancient Mongolian horses from archaeological cultures of different periods indicates that some ancient mitotypes were present in Mongolian horse populations for several thousand years. The geographic origins of the Iranian horse and of the aforementioned Arabian horse are the same: they originated in the Middle East region. Taking into account all the above information, we can infer that the clade with ancient Mongolian horses Er-1 and Gan-11 mainly includes horses originating in Asia.

The haplotype of ancient horse Gan14 from the Ganga Tsagaan ereg archaeological site belongs to haplogroup L and forms a clade with modern horses of the Middle East (an unspecified Iranian breed) and Chinese horse breeds. This clade contains horses of the Chinese Yunnan pony and Tengchong breeds, which belong to the southwest Chinese group of horse breeds: the Yunnan Sichuan horses. Horses of this group are small, short, and slender and originate from the mountainous areas of southwest China; phenotypically, they are very different from the Tibetan horse and Northern China horse (including the Mongolian horse) [45,48,49]. Another study, involving an analysis of protein polymorphisms, has revealed that the Mongolian horse is the breed ancestral to the Yunnan horse [3].

The analysis of the mtDNA control region sequences has also confirmed the hypothesis that domestic horses of China descended from both imported (outside of China) and local horses, whereas the imported horses were introduced from the northern regions of the Eurasian steppe [50].

The ancient horse Gan18 studied here is genetically closest to the modern horse of Chinese breed Jinjiang (both belong to haplogroup L). The Jinjiang horse is a unique modern breed distributed in the coastal areas of Central and South East China and is most genetically and phenotypically distinct from other southern horse breeds of China [48,49,51]. Jinjiang horses have muscular bodies and the draft-type conformation and exhibit specific adaptations to the hot and humid conditions of the south coastal areas, whereas other southern horse breeds are usually slim and dwarflike. Despite these phenotypic differences, the Yunnan and Jinjiang horses are genetically closely related, as proven by analyzing the whole-genome SNPs [51].

From the above information, it can be concluded that the ancient horse Gan14 studied here shares ancestry with modern horses of the Iranian and the Yunnan horse breeds, which are phenotypically different from the Tibetan and the Mongolian horses. Horse Gan18 is most closely related to the Jinjiang horse, which phenotypically differs from Yunnan horses, but they are closely related to each other. Taking into account the above consideration, it can be supposed that the mitotypes of Chinese horses related to the ancient Mongolian ones originate from Mongolia.

Two ancient horses, Gan1 and Gan3 from the Ganga Tsagaan ereg archaeological site, belong to mitogroup M and are especially close to one of its mitotypes comprising the modern horse of the Caspian pony breed. The Caspian pony is one of the oldest breeds with origins in the Middle East and was most likely developed on the territory of northern Iran during the Achaemenid Empire ~2.5–3.0 thousand years ago. Caspian ponies are horses of small stature and have probably resulted from natural hybridization between *Equus ferus caballus* and *Equus ferus przewalskii* [52]. However, this hypothesis has not yet been proven. Based on the analysis of a genome-wide set of autosomal SNPs, it has been determined that the Caspian horse falls into a clade with the other Middle Eastern breeds, the Arabian and the Akhal-Teke [1]. It can be assumed that the ancient horses of the Ganga Tsagaan ereg site are closely related to some Caspian horses of Middle Eastern origin.

4.2. Cultural Context

The Mongolian horse is one of the most ancient breeds, is relatively unmanaged, and has existed on the same territory for thousands of years [1], and the mitotypes closely related to the ancient Mongolian ones belong to modern horses of the breeds that—according to the literature data—are quite ancient (several thousand years) and indigenous [42,43,45,47–49,51,52]. Therefore, one can suggest possible migration routes of the obtained ancient mitotypes. However, due to the limited dataset, further research on ancient DNA of horses belonging to earlier archaeological cultures is needed for more accurate determination of these migration pathways.

The five studied ancient horses (Ganga Tsagaan ereg site) belong to the Khereksur and Deer Stone nomadic culture of Mongolia. The use of horses and camels provided mobility and some military superiority for the nomads over the farmers of Eurasia in the preindustrial era [53]. The widespread use of ritual horse sacrifices, typical of this culture, occurred at the same time as the spread of horse riding, horse milk utilization, and a significant increase in the proportion of horse meat in the human diet in most of Central Asia [54]. The Mongolian tribes of the Khereksur and Deer Stone culture interacted with the population of other regions, including remote ones (China, the Middle East, and North Asia) [16]. A recent study, based on an analysis of complete mitochondrial genome sequences, showed that the Ushkin-Uver archaeological site horses (9th to mid-8th century BC) of the same culture are closely related to earlier Chalcolithic Botai culture horses (mid-4th millennium BC) of the North Asian steppe and horses of Moldova (15th to 11th century BC, Mciurin site) and Germany (late 3rd to mid-2nd millennium BC, Schloßvippach site) of the Bronze Age [4]. On the basis of the above information, it cannot be presumed that the mitotypes of the Ushkin-Uver horses are ancestral for related mitotypes, especially because a complete nuclear genome analysis has revealed the origin of these horses from the horses of the South Ural (Russia) Sintashta culture of the Bronze Age (2nd millennium BC) [4]. The Ganga Tsagaan ereg horses belong to an earlier time period (late 12th to mid-10th century BC) than do Ushkin-Uver horses (9th to mid-8th century BC). We did not use the mitogenome consensus sequences of Ushkin-Uver ancient horses in the phylogeographic analysis because they had not been uploaded to the nucleotide databases. If we take into account the data obtained from the Ushkin-Uver horses, it does not seem correct to draw final conclusions about the ancestry of ancient Ganga Tsagaan ereg horse mitotypes toward the similar mitotypes of modern horses because it is likely that they were introduced to Mongolia in an earlier epoch.

Given that we found some continuity between the Mongolian horse populations of the Khereksur and Deer Stone culture and the Xiongnu culture, the occurrence of related mitotypes among the horse populations of the native breeds of China, the Middle East, and Italy can be explained by the relationship between the Xiongnu Empire and the Han Empire of China as well as the Roman empire. Some historical and archaeological data also suggest the continuity between the Khereksur and Deer Stone culture and the Xiongnu culture. However, which archaeological culture is ancestral in relation to the Xiongnu has not been identified yet [18]. What unites the cultures of the Khereksur and Deer Stone and the Xiongnu is the ritual burial of heads and hooves of horses, sheep, goats, cattle, camels, and dogs [54].

Much historical and archaeological evidence indicates that horse domestication took place in Central Asia, from where domesticated horses were distributed to other regions. Their introgression with wild horse populations in Mongolia and subsequent migration southward to China at the end of the 2nd millennium BC is assumed [55]. Around the 12th century BC, during the existence of the Chinese Shang state, when the Khereksur and Deer Stone culture was being formed on the territory of Mongolia, domesticated horses and horse-drawn carriages became widespread throughout Central China, as evidenced by written sources and the presence of horse bones in the corresponding burials [56]. In the next epoch of the Western Zhou (late 11th to early 8th century BC), there are reports of military clashes with the armed cavalry of the northern nomads. At all stages of the relationship between the Xiongnu Empire and the Han Empire (war, peaceful coexistence, including cross-border trade, gift giving, and payment of tributes), horses were exchanged, and a larger number of horses were exported from the Xiongnu to the Han Empire, not vice versa [18]. However, in the 3rd to 2nd century BC, the Chinese Empire did not have enough horses for its army. To increase the mobility of the troops, the Chinese people began to breed horses brought by Chinese merchants and travelers from the Fergana Valley [57–59]. From the above information, we can conclude that there are several distribution pathways of the horse mitotypes being considered: disengaged migration of ancient horses from Central Asia to China and Mongolia, migration to Mongolia via China, or to China via Mongolia. It is difficult to find out which of the considered hypotheses is the most correct based only on the data available.

Archaeological evidence also confirms the existence of long-distance contacts between the Xiongnu Empire and the Roman Empire. In the elite burial mound of the Noin-Ula site in Northern Mongolia, an antique silver plate has been unearthed portraying goddess Artemis and a satyr; this plate must have been one of horse harness parts [60]. Furthermore, a cup made of Roman glass has been discovered in the Xiongnu cemetery Gol Mod 2 of Northern Mongolia [61]. According to results of a polymorphism analysis of mitochondrial D-loop hypervariable region 1 (HV1), it has been determined that a buried person from a large settlement of Roman province Vagnari in Southern Italy belongs to a haplogroup originating from East Asia [62]. Findings in several studies reveal close genetic relationships between the mitotypes of Xiongnu culture ancient horses and indigenous horses of the Mediterranean region [4,63]. A molecular genetic study analyzing partial sequences of mitogenomes has shown a close phylogenetic relationship between ancient horses of Xiongnu royal tomb complex Tsaram (the Republic of Buryatia, Russia) and a modern horse of native Italian breed Maremmano [63], which was developed by the Etruscans in the 8th century BC [64]. An analysis of whole-mitogenome sequences has revealed that some Xiongnu horses were related to the horses from the territory of France dating to the Gallo-Roman period of classical antiquity (site of "Boulevard de la Courtille C277") [4]. The consensus sequences of ancient horse mitogenomes obtained in the above studies were not used in our phylogeographic analysis because some sequences represent only a partial mitochondrial genome, while others had not been uploaded to the nucleotide databases.

The existence of related mitotypes in the ancient Mongolian horse populations of the Khereksur and Deer Stone culture and the Xiongnu culture indicates certain continuity of the horse populations of these cultures. The presence of similar mitotypes in the modern horse populations of breeds indigenous to Eastern and Central Asia, the Middle East, and the Mediterranean region may reflect migration routes of the ancient horse populations associated with various contacts between the Xiongnu Empire, Han Empire, and Roman Empire. Archaeological, historical, and molecular genetic data indicate that the direction of domesticated-horse mitotype migration is likely to have been from Central Asia to other regions. Nonetheless, to identify the directions more precisely, it is necessary to conduct genetic studies involving a larger number of ancient horse samples of various archaeological cultures and to include nuclear markers.

4.3. Population Dynamics of the Mongolian Horse Breed

As shown in the constructed phylogenetic tree (Figure 1), the majority of modern and historical (20th century) horses of the Mongolian breed are affiliated with the same haplogroups as the ancient Mongolian horses studied here. Only one modern Mongolian horse was found to be exceptional and belongs to haplogroup A, and two other modern horse haplotypes are located in distant tree branches and diverged from other haplogroups 70–145 thousand years ago (median age estimates). The ancient Mongolian horse populations being investigated were found not to possess these basal haplotypes, which seems unusual. Although several studies involving an analysis of the mitogenome hypervariable-region sequences have also revealed higher mitochondrial diversity of modern Mongolian horse populations, the following mitotypes have been identified in ancient populations: A, D2, E, and X2b [12,13], while in modern ones: B2, D3, I, K, K2a, K2b, K3, X2, X3, and X3c1, according to the classification of Cieslak et al. [12]. This result may be due to a larger sample set size of the modern horses used in the research studies. To determine the actual genetic diversity of ancient Mongolian horse populations, further research on ancient DNA including whole-genome analysis is needed.

5. Conclusions

Our results revealed some continuity in a mitochondrial haplotype pattern between the Khereksur and Deer Stone Mongolian horse population and the Xiongnu Mongolian horse population. In the context of possible migration routes, we determined the mitotype kinship of the investigated ancient Mongolian horses and of the modern indigenous ones from Eastern and Central Asia, the Middle East, and the Mediterranean region. For more correct assessments of genetic diversity, continuity, and migration directions of horse populations from different archaeological cultures, further ancient-DNA studies are needed.

Supplementary Materials: The following are available online at https://www.mdpi.com/2073-4425/12/3/412/s1, Figure S1: Ancient DNA damage patterns, Figure S2: Read length distribution plot for the ancient sample libraries, Figure S3: Visualization of the coverage depth of the obtained mitogenomes of ancient Mongolian horses, Figure S4: Initial phylogenetic tree (with *Equus asinus*) built here showing the 95% height posterior density of age estimates, Figure S5: Initial phylogenetic tree (without *Equus asinus*) built here showing the 95% height posterior density of age estimates,

Table S1: Information about bone specimens of the studied ancient horses, Table S2: Information about the mitogenome sequences from GenBank.

Author Contributions: Project administration, A.S.G. and A.A.T.; supervision, N.V.V., A.A.T., and A.S.G.; conceptualization, N.V.V. and A.A.T.; funding acquisition, A.A.T. and N.V.V.; resources, A.A.T. and T.-O.I.; investigation, M.A.K. and A.A.T.; methodology, N.V.V. and A.S.D.; data curation, M.A.K. and A.I.M.; formal analysis, M.A.K. and A.I.M.; validation, M.A.K.; visualization, M.A.K.; writing—original draft preparation, M.A.K.; writing—review and editing, M.A.K., N.V.V., A.A.T., A.I.M., and V.A.T. All authors have read and agreed to the published version of the manuscript.

Funding: This research was funded by the Russian Foundation for Basic Research, grant numbers 19-59-15001 and 20-04-00213.

Institutional Review Board Statement: The study was approved by the Ethics Committee on Animal and Human Research at the Institute of Molecular and Cellular Biology, Russia (permit No. 01/20 of 11 February 2020).

Informed Consent Statement: Informed consent was obtained from all subjects involved in the study.

Data Availability Statement: The data presented in this study are openly available in [Sequence Read Archive database (SRA), GenBank database], reference numbers [PRJNA694234, MW534078, MW534079, MW534080, MW534081, MW534082, MW534083].

Acknowledgments: The authors gratefully acknowledge the resources provided by the "Molecular and Cellular Biology" core facility of the IMCB SB RAS. We appreciate the help of Olga N. Kusliy and Nikolai A. Shevchuk with editing the English language and style of the article.

Conflicts of Interest: The authors declare no conflict of interest. The funders had no role in the design of the study; in the collection, analyses, or interpretation of the data; in the writing of the manuscript; or in the decision to publish the results.

References

1. Petersen, J.L.; Mickelson, J.R.; Cothran, E.G.; Andersson, L.S.; Axelsson, J.; Bailey, E.; Bannasch, D.; Binns, M.M.; Borges, A.S.; Brama, P.; et al. Genetic Diversity in the Modern Horse Illustrated from Genome-Wide SNP Data. *PLoS ONE* **2013**, *8*, e54997. [CrossRef]
2. Hendricks, B.L. *International Encyclopedia of Horse Breeds*; The University of Oklahoma Press, Publishing Division of the University: Norman, OK, USA, 1995.
3. Nozawa, K.; Shotake, T.; Ito, S.; Kawamoto, Y. Phylogenetic Relationships among Japanese Native and Alien Horses Estimated by Protein Polymorphisms. *J. Equine Sci.* **1998**, *9*, 53–69. [CrossRef]
4. Fages, A.; Hanghøj, K.; Khan, N.; Gaunitz, C.; Seguin-Orlando, A.; Leonardi, M.; McCrory Constantz, C.; Gamba, C.; Al-Rasheid, K.A.S.; Albizuri, S.; et al. Tracking Five Millennia of Horse Management with Extensive Ancient Genome Time Series. *Cell* **2019**, *177*, 1419–1435.e31. [CrossRef] [PubMed]
5. Gaunitz, C.; Fages, A.; Hanghøj, K.; Albrechtsen, A.; Khan, N.; Schubert, M.; Seguin-Orlando, A.; Owens, I.J.; Felkel, S.; Bignon-Lau, O.; et al. Ancient genomes revisit the ancestry of domestic and Przewalski's horses. *Science* **2018**, *360*, 111–114. [CrossRef] [PubMed]
6. Kim, K.-I.; Yang, Y.-H.; Lee, S.-S.; Park, C.; Ma, R.; Bouzat, J.L.; Lewin, H.A. Phylogenetic relationships of Cheju horses to other horse breeds as determined by mtDNA D-loop sequence polymorphism. *Anim. Genet.* **1999**, *30*, 102–108. [CrossRef]
7. Cho, G.J. Genetic Relationship among the Korean Native and Alien Horses Estimated by Microsatellite Polymorphism. *Asian-Australas. J. Anim. Sci.* **2006**, *19*, 784–788. [CrossRef]
8. Tozaki, T.; Takezaki, N.; Hasegawa, T.; Ishida, N.; Kurosawa, M.; Tomita, M.; Saitou, N.; Mukoyama, H. Microsatellite variation in Japanese and Asian horses and their phylogenetic relationship using a European horse outgroup. *J. Hered.* **2003**, *94*, 374–380. [CrossRef]
9. Kovalev, A.A.; Erdenebaatar, D.; Iderkhangai, T.-O. Discovery of new Bronze Age culture in the South of Mongolia. In *Proceedings of the Ancient Cultures of Mongolia and Baikal Siberia, Ulaanbaatar, Mongolia, 5–9 September 2012*; National University of Mongolia: Ulaanbaatar, Mongolia, 2012; pp. 175–182.
10. Kovalev, A.A.; Erdenebaatar, D.; Rukavishnikova, I.V. A ritual complex with deer stones at Uushigiin Uvur/Ulaan Uushig, Mongolia: Composition and construction stages (based on the 2013 excavations). *Archaeol. Ethnol. Anthropol. Eurasia* **2016**, *44*, 82–92. (In Russian) [CrossRef]
11. Youn, M.; Kim, J.C.; Kim, H.K.; Tumen, D.; Navaan, D.; Erdene, M. Dating the Tavan Tolgoi Site, Mongolia: Burials of the Nobility from Genghis Khan's Era. *Radiocarbon* **2007**, *49*, 685–691. [CrossRef]
12. Cieslak, M.; Pruvost, M.; Benecke, N.; Hofreiter, M.; Morales, A.; Reissmann, M.; Ludwig, A. Origin and History of Mitochondrial DNA Lineages in Domestic Horses. *PLoS ONE* **2010**, *5*, e15311. [CrossRef] [PubMed]

13. Pilipenko, A.S.; Romaschenko, A.G.; Molodin, V.I.; Parzinger, H.; Kobzev, V.F. Mitochondrial DNA studies of the Pazyryk people (4th to 3rd centuries BC) from northwestern Mongolia. *Archaeol. Anthropol. Sci.* **2010**, *2*, 231–236. [CrossRef]
14. Tishkin, A.A. "Deer" stones of Mongolia and adjacent territories as one of the indicators of the archaic nomadic empire (to the formulation of the question). In *Proceedings of the V (XXI) All-Russian Archaeological Congress, Belokurikha, Russia, 1–8 October 2017*; Altai State University: Barnaul, Russia, 2017; p. 1026.
15. Iderkhangai, T.-O.; Mijiddorj, E.; Orgilbayar, S.; Galbadrakh, B.; Erdene, J.; nrbayar, B.; Enkhmagnay, G. *2015 Archaeological Rescue Excavation Report on the Egiin Gol Hydroelectric Dam, Khutag Undur Sum, Bulgan Aimag*; Ulaanbaatar State University: Ulaanbaatar, Mongolia, 2015; pp. 67–150.
16. Taylor, W. Horse demography and use in Bronze Age Mongolia. *Quat. Int.* **2017**, *436*, 270–282. [CrossRef]
17. Taylor, W.T.T.; Jargalan, B.; Lowry, K.B.; Clark, J.; Tuvshinjargal, T.; Bayarsaikhan, J. A Bayesian chronology for early domestic horse use in the Eastern Steppe. *J. Archaeol. Sci.* **2017**, *81*, 49–58. [CrossRef]
18. Kradin, N.N. *The Xiongnu Empire*, 2nd ed.; "The Publishing Group Logos" Ltd.: Moscow, Russia, 2001; ISBN 5-94010-124-0.
19. Garutt, V.E.; Yuriev, K.B. Paleofauna of the Ivolginsky settlement according to archaeological excavations of 1949–1956. *Archaeol. Festschr.* **1959**, 80–82.
20. Rudenko, S.I. *The Culture of the Huns and the Noin-Ula Burial Mounds*; Academy of Sciences of the USSR: Moscow/Leningrad, Russia, 1962.
21. Yang, D.Y.; Eng, B.; Waye, J.S.; Dudar, J.C.; Saunders, S.R. Improved DNA extraction from ancient bones using silica-based spin columns. *Am. J. Phys. Anthropol.* **1998**, *105*, 539–543. [CrossRef]
22. Sanderson, C.; Radley, K.; Mayton, L. Ethylenediaminetetraacetic Acid in Ammonium Hydroxide for Reducing Decalcification Time. *Biotech. Histochem.* **1995**, *70*, 12–18. [CrossRef] [PubMed]
23. Maricic, T.; Whitten, M.; Pääbo, S. Multiplexed DNA Sequence Capture of Mitochondrial Genomes Using PCR Products. *PLoS ONE* **2010**, *5*, e14004. [CrossRef]
24. Vorobieva, N.V.; Makunin, A.I.; Druzhkova, A.S.; Kusliy, M.A.; Trifonov, V.A.; Popova, K.O.; Polosmak, N.V.; Molodin, V.I.; Vasiliev, S.K.; Shunkov, M.V.; et al. High genetic diversity of ancient horses from the Ukok Plateau. *PLoS ONE* **2020**, *15*, e0241997. [CrossRef] [PubMed]
25. Schubert, M.; Jónsson, H.; Chang, D.; Der Sarkissian, C.; Ermini, L.; Ginolhac, A.; Albrechtsen, A.; Dupanloup, I.; Foucal, A.; Petersen, B.; et al. Prehistoric genomes reveal the genetic foundation and cost of horse domestication. *Proc. Natl. Acad. Sci. USA* **2014**, *111*, E5661–E5669. [CrossRef] [PubMed]
26. Schubert, M.; Lindgreen, S.; Orlando, L. AdapterRemoval v2: Rapid adapter trimming, identification, and read merging. *BMC Res. Notes* **2016**, *9*, 88. [CrossRef] [PubMed]
27. Li, H.; Durbin, R. Fast and accurate short read alignment with Burrows-Wheeler transform. *Bioinformatics* **2009**, *25*, 1754–1760. [CrossRef] [PubMed]
28. Jónsson, H.; Ginolhac, A.; Schubert, M.; Johnson, P.L.F.; Orlando, L. mapDamage2.0: Fast approximate Bayesian estimates of ancient DNA damage parameters. *Bioinformatics* **2013**, *29*, 1682–1684. [CrossRef] [PubMed]
29. Katoh, K.; Misawa, K.; Kuma, K.; Miyata, T. MAFFT: A novel method for rapid multiple sequence alignment based on fast Fourier transform. *Nucleic Acids Res.* **2002**, *30*, 3059–3066. [CrossRef] [PubMed]
30. Lanfear, R.; Calcott, B.; Ho, S.Y.W.; Guindon, S. PartitionFinder: Combined Selection of Partitioning Schemes and Substitution Models for Phylogenetic Analyses. *Mol. Biol. Evol.* **2012**, *29*, 1695–1701. [CrossRef] [PubMed]
31. Suchard, M.A.; Lemey, P.; Baele, G.; Ayres, D.L.; Drummond, A.J.; Rambaut, A. Bayesian phylogenetic and phylodynamic data integration using BEAST 1.10. *Virus Evol.* **2018**, *4*, vey016. [CrossRef] [PubMed]
32. Ferreira, M.A.R.; Suchard, M.A. Bayesian analysis of elapsed times in continuous-time Markov chains. *Can. J. Stat.* **2008**, *36*, 355–368. [CrossRef]
33. Drummond, A.J.; Nicholls, G.K.; Rodrigo, A.G.; Solomon, W. Estimating mutation parameters, population history and genealogy simultaneously from temporally spaced sequence data. *Genetics* **2002**, *161*, 1307–1320.
34. Kingman, J.F.C. The coalescent. *Stoch. Process. Appl.* **1982**, *13*, 235–248. [CrossRef]
35. Rambaut, A.; Drummond, A.J.; Xie, D.; Baele, G.; Suchard, M.A. Posterior Summarization in Bayesian Phylogenetics Using Tracer 1.7. *Syst. Biol.* **2018**, *67*, 901–904. [CrossRef]
36. Drummond, A.J.; Rambaut, A. BEAST: Bayesian evolutionary analysis by sampling trees. *BMC Evol. Biol.* **2007**, *7*, 1–8. [CrossRef]
37. Sawyer, S.; Krause, J.; Guschanski, K.; Savolainen, V.; Pääbo, S. Temporal Patterns of Nucleotide Misincorporations and DNA Fragmentation in Ancient DNA. *PLoS ONE* **2012**, *7*, e34131. [CrossRef] [PubMed]
38. Achilli, A.; Olivieri, A.; Soares, P.; Lancioni, H.; Kashani, B.H.; Perego, U.A.; Nergadze, S.G.; Carossa, V.; Santagostino, M.; Capomaccio, S.; et al. Mitochondrial genomes from modern horses reveal the major haplogroups that underwent domestication. *Proc. Natl. Acad. Sci. USA* **2012**, *109*, 2449–2454. [CrossRef] [PubMed]
39. Orlando, L.; Ginolhac, A.; Zhang, G.; Froese, D.; Albrechtsen, A.; Stiller, M.; Schubert, M.; Cappellini, E.; Petersen, B.; Moltke, I.; et al. Recalibrating Equus evolution using the genome sequence of an early Middle Pleistocene horse. *Nature* **2013**, *499*, 74–78. [CrossRef] [PubMed]
40. Librado, P.; Der Sarkissian, C.; Ermini, L.; Schubert, M.; Jónsson, H.; Albrechtsen, A.; Fumagalli, M.; Yang, M.A.; Gamba, C.; Seguin-Orlando, A.; et al. Tracking the origins of Yakutian horses and the genetic basis for their fast adaptation to subarctic environments. *Proc. Natl. Acad. Sci. USA* **2015**, *112*, E6889–E6897. [CrossRef] [PubMed]

41. Boeskorov, G.G.; Potapova, O.R.; Protopopov, A.V.; Plotnikov, V.V.; Maschenko, E.N.; Shchelchkova, M.V.; Petrova, E.A.; Kowalczyk, R.; van der Plicht, J.; Tikhonov, A.N. A study of a frozen mummy of a wild horse from the Holocene of Yakutia, East Siberia, Russia. *Mammal. Res.* **2018**, *63*, 307–314. [CrossRef]
42. Cosgrove, E.J.; Sadeghi, R.; Schlamp, F.; Holl, H.M.; Moradi-Shahrbabak, M.; Miraei-Ashtiani, S.R.; Abdalla, S.; Shykind, B.; Troedsson, M.; Stefaniuk-Szmukier, M.; et al. Genome Diversity and the Origin of the Arabian Horse. *Sci. Rep.* **2020**, *10*, 9702. [CrossRef]
43. Szontagh, A.; Bán, B.; Bodó, I.; Cothran, E.G.; Hecker, W.; Józsa, C.; Major, Á. Genetic diversity of the Akhal-Teke horse breed in Turkmenistan based on microsatellite analysis. In *Conservation Genetics of Endangered Horse Breeds*; Bodo, I., Alderson, L., Langlois, B., Eds.; Wageningen Academic Publishers: Bled, Slovenia, 2005; pp. 123–128.
44. Kovalevskaya, V.B. Ancestors of the Oriental Horse in Eurasia: Origin and Distribution. *Archaeol. Ethnol. Anthropol. Eurasia* **2020**, *48*, 129–139. [CrossRef]
45. Yang, L.; Kong, X.; Yang, S.; Dong, X.; Yang, J.; Gou, X.; Zhang, H. Haplotype diversity in mitochondrial DNA reveals the multiple origins of Tibetan horse. *PLoS ONE* **2018**, *13*, e0201564. [CrossRef] [PubMed]
46. Du, D.; Zhao, C.; Zhang, H.; Han, G. Genetic Diversity of Tibetan Horse and its Relationships with Mongolian Horse and Ningqiang Pony Assessed by Microsatellite Polymorphism. *Asian J. Anim. Vet. Adv.* **2011**, *6*, 564–571. [CrossRef]
47. Morelli, L.; Useli, A.; Sanna, D.; Barbato, M.; Contu, D.; Pala, M.; Cancedda, M.; Francalacci, P. Mitochondrial DNA lineages of Italian Giara and Sarcidano horses. *Genet. Mol. Res.* **2014**, *13*, 8241–8257. [CrossRef] [PubMed]
48. Ling, Y.; Ma, Y.; Guan, W.; Cheng, Y.; Wang, Y.; Han, J.; Jin, D.; Mang, L.; Mahmut, H. Identification of Y Chromosome Genetic Variations in Chinese Indigenous Horse Breeds. *J. Hered.* **2010**, *101*, 639–643. [CrossRef] [PubMed]
49. Ling, Y.H.; Ma, Y.H.; Guan, W.J.; Cheng, Y.J.; Wang, Y.P.; Han, J.L.; Mang, L.; Zhao, Q.J.; He, X.H.; Pu, Y.B.; et al. Evaluation of the genetic diversity and population structure of Chinese indigenous horse breeds using 27 microsatellite markers. *Anim. Genet.* **2011**, *42*, 56–65. [CrossRef] [PubMed]
50. Yang, Y.; Zhu, Q.; Liu, S.; Zhao, C.; Wu, C. The origin of Chinese domestic horses revealed with novel mtDNA variants. *Anim. Sci. J.* **2017**, *88*, 19–26. [CrossRef] [PubMed]
51. Ma, H.; Wang, S.; Zeng, G.; Guo, J.; Guo, M.; Dong, X.; Hua, G.; Liu, Y.; Wang, M.; Ling, Y.; et al. The Origin of a Coastal Indigenous Horse Breed in China Revealed by Genome-Wide SNP Data. *Genes* **2019**, *10*, 241. [CrossRef]
52. Seyedabadi, H.R.; Amirinia, S.; Bana, B.M.H.; Emrani, H. Parentage verification of Iranian Caspian horse using microsatellites markers. *Iran. J. Biotechnol.* **2006**, *4*, 260–264.
53. Khazanov, A.M. Ecological limitations of nomadism in the Eurasian steppes and their social and cultural implications. *Asian Afr. Stud.* **1990**, *24*, 1–15.
54. Taylor, W.; Fantoni, M.; Marchina, C.; Lepetz, S.; Bayarsaikhan, J.; Houle, J.-L.; Pham, V.; Fitzhugh, W. Horse sacrifice and butchery in Bronze Age Mongolia. *J. Archaeol. Sci. Rep.* **2020**, *31*, 102313. [CrossRef]
55. Honeychurch, W. *Inner Asia and the Spatial Politics of Empire*; Springer: New York, NY, USA, 2015; ISBN 978-1-4939-1814-0.
56. Kelekna, P. *The Horse in Human History*; Cambridge University Press: Cambridge, UK, 2009; ISBN 978-0-521-51659-4.
57. Creel, H.G. The role of the horse in Chinese history. *Am. Hist. Rev.* **1965**, *70*, 647–672. [CrossRef]
58. Panov, V.A. *To the History of the Peoples of Central Asia*, 2nd ed.; A. V. Dattan: Vladivostok, Russia, 1918.
59. Schafer, E.H. *The Golden Peaches of Samarkand*; Nauka: Moscow, Russia, 1981.
60. Polosmak, N.V. The light of distant Hellas. *Sci. First Hand* **2011**, *30*, 88–101.
61. Erdenebaatar, D.; Iderkhangai, T.; Galbadrakh, B.; Minjiddorj, E.; Orgilbayar, S. Excavations of satellite burial 30, tomb 1 complex, Gol Mod 2 necropolis. In *Proceedings of the Xiongnu archaeology: Multidisciplinary perspectives of the first steppe empire in Inner Asia*; Brosseder, U., Miller, B.K., Eds.; Rheinische Friedrich-Wilhelms-Universitat: Ulaanbaatar, Mongolia, 2011; pp. 303–314.
62. Prowse, T.L.; Barta, J.L.; von Hunnius, T.E.; Small, A.M. Stable isotope and mitochondrial DNA evidence for geographic origins on a Roman estate at Vagnari (Italy). *J. Rom. Archaeol.* **2010**, *78*, 175–198.
63. Kusliy, M.A.; Druzhkova, A.S.; Popova, K.O.; Vorobieva, N.V.; Makunin, A.I.; Yurlova, A.A.; Tishkin, A.A.; Minyaev, S.S.; Trifonov, V.A.; Graphodatsky, A.S.; et al. Genotyping and coat colour detection of ancient horses from Buryatia. *Tsitologiia* **2016**, *58*, 304–308. [PubMed]
64. Giontella, A.; Cardinali, I.; Lancioni, H.; Giovannini, S.; Pieramati, C.; Silvestrelli, M.; Sarti, F.M. Mitochondrial DNA Survey Reveals the Lack of Accuracy in Maremmano Horse Studbook Records. *Animals* **2020**, *10*, 839. [CrossRef] [PubMed]

Article

Mitochondrial DNA Profiles of Individuals from a 12th Century Necropolis in Feldioara (Transylvania)

Alexandra Gînguță [1], Ioana Rusu [1,2,*], Cristina Mircea [1,2], Adrian Ioniță [3], Horia L. Banciu [1,4] and Beatrice Kelemen [1,2]

1 Department of Molecular Biology and Biotechnology, Faculty of Biology and Geology, Babeș-Bolyai University, 400006 Cluj-Napoca, Romania; alexandra.ginguta@ubbcluj.ro (A.G.); cristina.mircea@ubbcluj.ro (C.M.); horia.banciu@ubbcluj.ro (H.L.B.); beatrice.kelemen@ubbcluj.ro (B.K.)
2 Molecular Biology Center, Interdisciplinary Research Institute on Bio-Nano-Sciences, Babeș-Bolyai University, 400271 Cluj-Napoca, Romania
3 Vasile Pârvan Institute of Archaeology, Romanian Academy, 010667 Bucharest, Romania; aionita67@yahoo.com
4 Centre for Systems Biology, Biodiversity, and Bioresources, Babeș-Bolyai University, 400006 Cluj-Napoca, Romania
* Correspondence: ioana.rusu@ubbcluj.ro

Abstract: The genetic signature of modern Europeans is the cumulated result of millennia of discrete small-scale exchanges between multiple distinct population groups that performed a repeated cycle of movement, settlement, and interactions with each other. In this study we aimed to highlight one such minute genetic cycle in a sea of genetic interactions by reconstructing part of the genetic story of the migration, settlement, interaction, and legacy of what is today the Transylvanian Saxon. The analysis of the mitochondrial DNA control region of 13 medieval individuals from Feldioara necropolis (Transylvania region, Romania) reveals a genetically heterogeneous group where all identified haplotypes are different. Most of the perceived maternal lineages are of Western Eurasian origin, except for the Central Asiatic haplogroup C seen in only one sample. Comparisons with historical and modern populations describe the contribution of the investigated Saxon settlers to the genetic history of this part of Europe.

Keywords: mitochondrial DNA; medieval individuals; Transylvania; population genetics

Citation: Gînguță, A.; Rusu, I.; Mircea, C.; Ioniță, A.; Banciu, H.L.; Kelemen, B. Mitochondrial DNA Profiles of Individuals from a 12th Century Necropolis in Feldioara (Transylvania). *Genes* **2021**, *12*, 436. https://doi.org/10.3390/genes12030436

Academic Editor: Antonio Amorim

Received: 25 January 2021
Accepted: 17 March 2021
Published: 19 March 2021

1. Introduction

The present-day territory of Romania reunites four distinct historical regions (Transylvania, Wallachia, Moldavia, and Dobruja), divided by the Carpathian Mountains Range that arches through the middle of the country. Until their unification into a single country in the 20th century, these provinces evolved rather independently, sometimes at odds with each other, for more than five centuries under different political and religious influences, being alternately challenged by distinct foreign powers [1]. This tangled history left its mark on the material culture, social organization, economic aspects, and genetic makeup of Romania, making it a space of convergence between different cultures. The consequences of past migrations, local admixture, and colonization events on the gene pool of current occupants of Romania's territory have been addressed in a few studies that aimed to uncover their mitochondrial DNA (mtDNA) diversity [2–4]. Among these, one study assessing the overall mtDNA control region variation has indicated that the maternal genetic landscape is relatively homogenous across Romania [3]. In a different study, it has been proposed that the Romanian Carpathians do not act as a strong genetic barrier, as the mtDNA haplogroup distribution of modern populations separated by this natural boundary showed only a small degree of genetic differentiation [4]. A more detailed comparison of mtDNA variation between the historical regions pointed out that the topology of the Romanian territory might have influenced past migration routes. The present Transylvanian population stands out by being closely related to its Central European counterpart, while the populations of provinces outside the Carpathian Arch were more similar to those in the Balkans [2].

Representing, today, the central-western region of Romania, Transylvania was, in the Middle Ages, a province with a special identity within the Kingdom of Hungary, whose structure was gradually integrated during the 10th–13th centuries. This integration was carried out in stages and involved mainly two populations, known in historiography as the Szeklers and the Saxons. The first group, already existing in the kingdom since the 10th century, had a predominantly military role in defending the borders [5]. The second group arrived in Central and Eastern European territories from the Occident as part of a huge wave of colonization starting in the middle of the 12th century. Most of the Western settlers that colonized southern and southeastern Transylvania came from Flanders and Saxony, and are identified today by the generic name of Saxons [6]. This term does not designate their precise origin; it is rather used as a conventional denomination for the privileged status of the settlers from German-speaking territories. In a second stage, the Teutonic Order settled in the south-eastern corner of Transylvania [7], a region which appears under the name of Ţara Bârsei (Burzenland) and whose boundaries were established on that occasion. This land endowment, made in a complex context, had the stated purpose of protecting the southeastern borders of the kingdom against the incursions of the Cumans. In fact, it is considered that the region where these settlers arrived was less than desirable for settlement precisely because of the proximity to the Turkic nomads—the Pechenegs and Cumans. The endowment was retracted 14 years later, in 1225, as a result of the king's dissatisfaction with the Order's activities and especially with their tendency to act independently and to acquiesce more to the Pope than to the king. Feldioara, whose German name is Marienburg, has gone down in history mainly as the seat of the Teutonic Knights in their short Transylvanian incursion. Located approximately in the center of the territory granted to this military order, Feldioara has been founded in the middle of the 12th century by a community of German settlers (peasants and craftsmen), in a place with an important strategic position.

These colonization events imprinted the demographic history of populations from the current territory of Transylvania creating a cultural complex space, but the extent of these influences and other ancestral fine details remain largely unknown. Few studies have explored the genetic features of ethnic groups presently living in Romania in the hope of a better understanding of their biological and cultural identity, connections to other populations, and genetic legacy [8–10]. Most of these sought to investigate the genetic diversity of modern Hungarian-speaking groups from a maternal perspective [9], Y-chromosomal haplotypes [11], or even based on genome-wide autosomal marker data [12]. The genetic data available for modern Transylvanian Saxons is even scarcer, being limited to a single study [13]. Using Y-STR haplotype data for 59 males, it was revealed that they are genetically closer to Germans, Romanians, and Transylvanian Székely than to other European populations.

The purpose of this study is to investigate the maternal (mitochondrial) genetic diversity of medieval individuals discovered in Feldioara necropolis, potential members or descendants of a group that attained the farthest German colonization limit [14], and to consider their similarities, differences, and links to their contemporary and succeeding populations.

2. Materials and Methods

2.1. Sample Background

The human archaeological remains analyzed in this study were discovered at Feldioara necropolis, also known as Marienburg (Brașov County). The cemetery was located in a square between the evangelical church and the parsonage. The excavations in this area were conducted between 1991 and 1995 [14], in 1998, and between 2006 and 2007 [15]. A number of 127 graves with a total of 145 burials were uncovered, 109 graves outside, and 18 inside the churchyard. Of the 127 investigated graves, most were simple graves, but also multiple graves (13 double, 2 triple, and one containing the remains of four dead) were found. Of these, 14 individuals were selected for genetic analysis based on the preservation status of the osteological material (Supplementary Figure S1). The age and sex of the selected individuals were previously determined in [15]: 13 adults (6 females, 6 males, and 1 indeterminate) and 1 subadult, who had an arrowhead between the ribs on the left side of the body, presumably the cause of death [16].

The routine funerary ritual in this necropolis involved the direct deposition of the adult bodies, without coffins, in graves with a step. This type of grave consists of a first larger rectangular hole continued by a smaller inferior pit outlining the shape and size of the human body, including a niche for the head. The funeral inventory is very poor, consisting of reused Roman coins, six Hungarian anonymous denars, and three lockrings with "S" -shaped ends [15]. While this kind of lockrings are found in a larger area (from Germany [17] to the Carpathian Basin [18] in the Bjelo Brdo's cultural space [19]) and were found in only two graves (M86, M113) in this necropolis, the anonymous Hungarian denars are minted only in the second half of the 12th century and remained in use usually for a short period of time due to frequent reminting. The six anonymous Hungarian denars found as grave goods in the Feldioara cemetery are attributed by numismatists to the reigns of Geza II (1141–1161) and Stephan III (1162–1172) and later coins were absent from the funerary inventory [14]. The beginning of use of this necropolis corresponds with the agreed start of Saxon settlement in Transylvania in the reign of Geza II and ends with the arrival of the Teutonic Knights in 1211 [14]. The stratigraphy observed during the excavation showed a layer of construction rubble covering the upper layer of the graves, originating probably in a massive edifice attributed to the Knights. Grave fills show no traces of construction debris. It is inferred that the burials at this site have been used for 2–3 generations, a fact supported by overlapping of two graves in five instances and three graves in two cases [14]. The inventory and analogies with the cemeteries of Western Europe and with those from Transylvania (all located in the Saxon colonization zone) show that the Feldioara cemetery belonged to the first wave of German settlers who arrived in Transylvania after the middle of the 12th century [14].

2.2. Ancient DNA Analysis

All DNA extractions and amplifications of the selected archaeological samples were performed in a sterile environment in laboratories exclusively dedicated to ancient DNA research at the Interdisciplinary Research Institute on Bio-Nano-Sciences, "Babeș-Bolyai" University, Romania. All steps were conducted following precise guidelines and appropriate workflow protocols for aDNA processing previously described [20]. Even though the persons who manipulated the samples during molecular analysis wore clean protective equipment to avoid contamination, their mtDNA signature was determined (Supplementary Table S1) and compared to the results obtained from the ancient samples.

The biological material used for DNA isolation was represented by dental samples with a very good preservation status for each tested historical individual. When available, multirooted teeth were selected. Prior to DNA extraction, each sample was decontaminated by mechanical removal of the external layer of the sample using a dental micro-drill (Marathon-3 Champion, Saeyang Microtech, Korea). Then, the samples were UV-irradiated (254 nm) for 45 min. Routinely, 100 mg of powder was collected from the dentine of the tooth root and then used for aDNA isolation. A negative control (no dental powder) was processed for every batch of three samples during the extraction and DNA amplification steps to monitor potential contamination with exogenous DNA.

The DNA extraction was performed in a designated pre-PCR area following a published silica-based protocol [21], but we adapted the described binding apparatus by using silica spin columns provided with the QIAquick PCR purification kit (Qiagen, Hilden, Germany) to allow for the recovery of products larger than 100 bp. In total, two 50 µl-TET buffer (10 mM Tris-HCL, 1 mM EDTA, 0.05% Tween-20, pH 8.0) aliquots were used for elution. The results were reinforced by at least two extractions or amplifications for several individuals, detailed in Supplementary Table S2.

The amplification of both hypervariable regions (HVR-I and HVR-II) of the human mitochondrial genome was performed using a set of eight mini-primer pairs, specially designed for highly degraded samples [22]. Besides the control region of the mitochondrial genome, three phylogenetic significant SNPs of the coding region (7028, 12308, 14766) were targeted to clarify the haplogroup classification of some samples as detailed in Supplementary Table S2, using the primers described in [23]. The amplification reactions of the targeted regions were set up in a final volume of 25 μL. Each PCR reaction consisted of 1.25 U MangoTaq™ DNA Polymerase (Bioline Reagents Ltd., London, UK), 1× MangoTaq™ Colored Reaction Buffer, 2.5 mM MgCl2, 0.2 mM dNTP mix, 0.3 μM each primer, 1–5 μL DNA extract, 0.5 μg/ μL BSA and PCR-grade water (Jena Bioscience, Jena, Germany) up to the final volume. The following cycle conditions were applied: initial denaturation at 95 °C for 5 min, 35 amplification cycles consisting of three steps (denaturation at 95 °C for 30 s, annealing at 46–56 °C depending on the primer pair, for 30 s and extension at 72 °C for 30 s) and final elongation at 72 °C for 5 min. A negative PCR control was included in each run. The PCR products were visualized on a 1.5% agarose electrophoresis gel and then purified using FavorPrep GEL and PCR Purification Kit (Favorgen Biotech Corp., Pingtung, Taiwan), according to the manufacturer's instructions. The fragments of interest were cloned using the Sticky-End Cloning Protocol available with the CloneJet PCR Cloning Kit (Thermo Scientific, Waltham, USA) and then a batch of six clones per amplicon were Sanger-sequenced at Macrogen Europe (Amsterdam, The Netherlands) using the standard primer pJET1.2R. If these first results were inconclusive another batch of six clones per PCR fragment were generated from an independent PCR reaction. All resulting sequences were aligned using the ClustalW algorithm for multiple sequences included in BioEdit Sequence Alignment Editor v. 7.2.5.0. The mtDNA sequence polymorphisms in the nucleotide position range 57 to 357 and 16,009 to 16,390 were identified relative to the revised Cambridge Reference Sequence (rCRS, NC_012920) [24]. Haplogroup determination was carried out according to the mtDNA phylogeny of PhyloTree build 17 using the HaploGrep2 web application [25,26]. All mitochondrial control region sequences generated during this study were submitted to GenBank under the accession numbers MW508518–MW508530. The mtDNA profiles of the investigated samples are reported in Table S3.

2.3. Population Genetic Analyses

The mtDNA data generated for the individuals sampled in this study were compared to ancient and modern mtDNA sequences deposited in online databases to identify their genetic relatedness. This comparison aims to provide traces of evidence on past migratory routes and geographic origin of the analyzed population. The human mitochondrial sequences retrieved from publicly available data, detailed below, were first manually grouped based on their age (medieval and contemporary), and then according to their geographic origin. The resulting datasets, one consisting of 21 historical populations (including the population considered in this study) and one, including 35 modern Eurasian populations, were used for haplogroup-based analysis, as well as for sequence-based analysis as detailed below. The population from Feldioara were compared to an ancient dataset consisting of 747 sequences of European populations and a Byzantine group [27]: Lombards from Italy [28,29] and Hungary [29], Avars [30–32], Vikings from Norway [33] and Denmark [34], medieval Basques [35], Italians [36], Bulgarians [37], medieval population of Conquest period from Hungary [30,38], medieval populations from Poland [39], Slovakia [40], Iceland [41], southeastern Romania [20,42] and Bavaria [43]. In addition to these medieval groups, an Iron Age population attributed to Goths [44] and a population from Italy dated to the Roman period [45] were used in comparative analyses. Characterizing spatio-temporal variables and mitochondrial haplogroup frequencies are given in Table S4. For comparison with modern Eurasian diversity, we used the mtDNA haplogroup frequencies and HVR-I sequences previously compiled in Rusu et al. [20] and the references therein.

Principal component analysis (PCA) was carried out based on mtDNA haplogroup frequencies of medieval and modern datasets. In total, 33 mitochondrial haplogroups were considered in the PCA of 21 ancient populations (Supplementary Table S4), whereas in the PCA with 35 present-day Eurasian populations 37 mtDNA haplogroups were considered (detailed in [20]). All PCAs were conducted using the prcomp function of the R stats package implemented in R 4.0.2 [46], and displayed in a two-dimensional space, showing the first two principal components (PC1 and PC2). A hierarchical clustering using Ward's agglomerative method [47] and Euclidean distance was employed on all genetic variations (PC) that resulted from the PCA of historical populations. The pvclust function in R [48] was used to calculate the probability values (*p*-values) for each cluster using 10,000 bootstrap replications. The significance of each cluster in the resulting dendrogram was given as an approximately unbiased (AU) *p*-value shown at branches.

Population comparisons using both modern and ancient reference sequences were estimated using Arlequin 3.5.2.2 [49]. Pairwise differences were calculated based on aligned HVR-I sequences of uniform sequence length, ranging from 16050 np to 16383 np (nucleotide position). The substitution model and the shape of Gamma parameter were assessed with ModelTest-NG [50]. For comparisons between the investigated population and both reference datasets, the Fst values were determined assuming a Tamura and Nei substitution model [51], 10,000 permutations and a significance level of 0.05. The linearized Slatkin Fst values and significant *p*-values of the historical populations were graphically represented as a level plot (Supplementary Table S5), while those of modern populations were visualized using multidimensional scaling (MDS) (Supplementary Table S6). The MDS plot was generated with metaMDS function based on Euclidean distances executed in the vegan library of R 4.0.2 [46,52].

3. Results and Discussion

3.1. *MtDNA Polymorphism*

We attempted to extract total DNA and subsequently amplify the two hypervariable regions of the human mitochondrial genome from 14 historical individuals from Feldioara. We identified the mtDNA polymorphism of the control region sequences spanning nucleotide positions 57–357 and 16009–16390 for all specimens, except one, labeled as M83a. Despite multiple extractions and amplifications for this sample with clean negative controls, the results were ambiguous, indicating potential contamination with exogenous DNA from the researcher performing the laboratory analysis. In this context, it seemed rational to exclude this sample from any further discussion and statistical analyses. The sequence motifs spanning the control region for the other 13 samples were assigned to mtDNA haplogroups in most cases with a very good overall quality of HaploGrep definitions (above 90%). The haplogroup affiliations were corroborated by diagnostic coding region sites which are all detailed in Table S3. Taking into consideration that equivalent results were obtained from replicate DNA extractions and amplifications following standard guidelines for ancient DNA analysis, as well as the detection of typical miscoding lesions in all cloned sequences support the trustworthiness of the generated mitochondrial data.

The maternal genetic composition of the analyzed group consisting of 13 samples indicates a high level of heterogeneity since it is comprised of 13 distinct haplotypes. The fact that these individuals do not share the same mtDNA profiles suggests that there are no strong maternal relationships among the analyzed group. The lack of identical maternal lineages might be a consequence of the small sample size, randomly selected from different parts of the necropolis, hardly any from adjacent burials (Supplementary Figure S1). However, other kinds of family connections cannot be ruled out solely based on the analysis of the uniparental marker that reveals the matrilineal perspective.

The mitochondrial haplogroup architecture of the analyzed medieval group includes predominantly common west Eurasian clades (H, HV, J, I, U4, and U5), but not exclusively, as one of the sequences belonging to a middle-aged female (M26) was classified into haplogroup C, an east Eurasian lineage (Figure 1). The genetic input from Central Asia has also been observed in medieval individuals from the province of Dobruja, but has been represented by a distinct mtDNA variant with a different phylogenetic history [20]. The maternal lineages originating from West Eurasia identified in the investigated population are different from those found in the small medieval group from southeastern Romania, except the common H2a2a, which may indicate different genetic influx in these geographic regions (Figure 1).

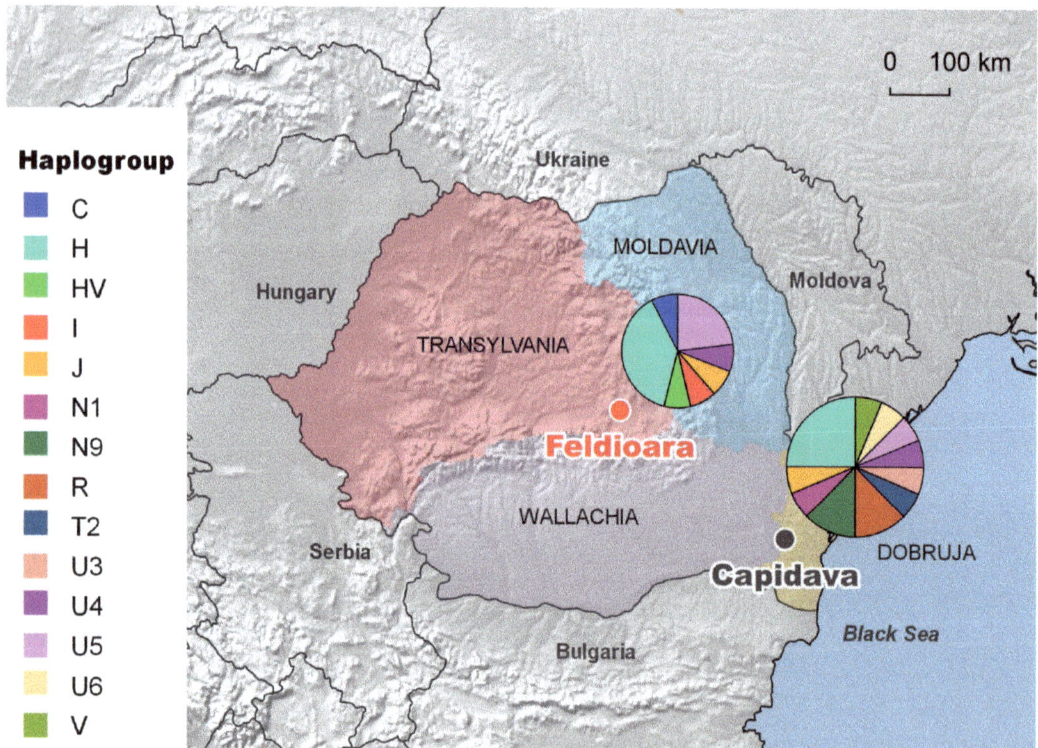

Figure 1. Geographic and genetic relationships between two medieval populations from the current territory of Romania: the population from Feldioara (labeled in red), located in the historical region of Transylvania, was investigated in this study, while the population from Capidava (labeled in grey, and labeled as seROU_med in further graphs and text), located in the province of Dobruja, was analyzed in previous studies [20,42]. Pie charts represent the frequencies of the major haplogroups in these populations. The size of the charts is proportional to the number of samples from each necropolis.

The sequence of the M26 sample showed a transition at 16357, characteristic of the C4a2 subclade. This variant is commonly found in the mtDNA pool of modern inhabitants of Siberia (particularly Evenks and Yakuts), while in Eastern Europe it is generally poorly represented [53,54]. This lineage and few derivatives were identified in ancient specimens such as: Iron Age Tagar individuals from Southern Siberia [55], Bronze Age representatives from the Baikal Lake region [56], as well as Bronze Age Kurgan specimens from the North Pontic Region [57]. C4a2a1 haplotype, distinguished by the 16171G mutation, has been identified in more recent samples dated to the medieval period, more specifically one individual from the Trans-Ural region and one attributed to Hungarian Conquerors that inhabited the Carpathian Basin during the tenth century [58]. The phylogeographic analysis of this mitochondrial lineage conducted in [58] pointed to the Kazakh or Baraba steppe as the most probable geographic origin. Based on this information and taking into account the historical context (Transylvania, which includes Feldioara, was gradually integrated into the Hungarian Kingdom during 10th–13th centuries), a feasible scenario is that this maternal lineage could have infiltrated in the Feldioara region along with the ancient Hungarians who migrated and conquered the Carpathian Basin in the 9th–10th centuries. The earliest occurrence of C4a2, until now, was documented in Early Bronze Ages samples from Shamanka (Baikal region) [59] and Middle Bronze Age samples associated with Deer Stone–Khirigsuur Complex from Mongolia [60], indicating that C4a2 most probably originated from the Mongolia–Baikal region.

Haplogroup H, the prevailing lineage in the modern gene pool of Europeans, accounts for nearly 40% of the mtDNA variation in the investigated group, being represented by different clades (H1a, H1b, H2a2a, H5, and H7f). For most of them, further classification into derivative sub-lineages might have been possible with the aid of whole coding region data, but not for H7f, which is precisely identified due to the presence of the diagnostic mutation 16168T in the reconstructed sequence of the M42a sample. This particular haplotype seems rather scarce in historical samples as it has been revealed only by the mitochondrial signature of a female from Isola Sacra associated with the Imperial Rome [61]. The M15 Feldioara sample contains defining mutations for HV9a, also identified in an individual of Viking Age from Norway, but which lacks the private mutation (16390A) present in our analyzed sample [33]. The HV9a lineage has been observed in the genomic data of Early Medieval individuals (about 500 AD.) with artificially deformed skulls discovered in present-day Bavaria, a genetically heterogeneous group comprising women that presumably originated from southeastern Europe in contrast to individuals with normal skulls from the same region that seemed to have mainly a northern and/or central European ancestry [43]. The second most abundant macro-haplogroup in the investigated population is U, with four samples classified into different subdivisions of this clade (U5b2a1a, U4a3a, and two distinct U5a1 haplotypes). Of these, U4a3a and U5a1d2a seem to be present only in few historical samples attributed to different cultures and time frames but mainly discovered in northern Europe (LM60mt—Middle Ages England [62], I2462—Early Bronze Age England [63], VK328—Viking Age Denmark [64], LP162mt—Early Medieval Denmark [62]). The genetic pattern of Feldioara sample M16a shows, besides a transition at position 207, a variant at 152 from the root node of haplogroup I, which leads to the affiliation of this individual to I2′3 lineage. A similar mtDNA profile has occurred in an ancient sample from the current territory of Germany associated with Unetice culture [65], a fact which implies a potential genetic connection between the investigated population and not only northern Europe, but also with the central part of the continent.

3.2. Population Genetic Analyses

To gain more clues regarding the genetic affinities between the sampled individuals as a group and other historical and modern populations reported in literature, we conducted haplogroup-based and sequenced-based analyses. The PCA of 21 ancient populations was constructed based on mtDNA haplogroup frequencies (Supplementary Table S4), and visually represented in a two-dimensional space (Figure 2a). The graph shows the rough clustering of most ancient European populations along the first two PCs, regardless of their geographic affiliation. Among them, only those medieval populations from southern Europe seem to be more tightly grouped. Even though the population dated to the Roman period originates from southern Italy it is separated from the South European cluster, likely due to its complex genetic background shaped by intense people interactions, imperial expansions and long-distance trades [61]. On PC1, the Avar elite group derived from the center of the Carpathian Basin (HUN_Avar-elite) is located in a marginal position. This is somehow expected as the gene pool of this group is dominated by a varied spectrum of Asian haplogroups [31]. Similarly, two isolated groups are revealed also on PC2 due to the contribution of N, R, and U suclades: a mid-Byzantine population from southwestern Anatolia (TUR_Byz) and a medieval population from southeastern Romania (seROU_med).

As a consequence, the medieval population from Transylvania (cROU-med) is distantly placed from the medieval population from Dobruja, indicating distinct genetic influences in these two regions separated by the Carpathian Arch. Comparable lines of evidence were drawn by the analysis of mtDNA variation in present-day Romanians [2]. The clear separation of the two medieval populations from distinct historical provinces of Romania is also revealed by the hierarchical clustering as they are located on different branches. The Ward type hierarchical clustering (Figure 2b) also shows the genetic associations of the studied population with historical sample sets from northern Europe (Vikings from Norway and Early medieval Icelanders), early medieval burials from Bavaria, as well as Longobards from Hungary.

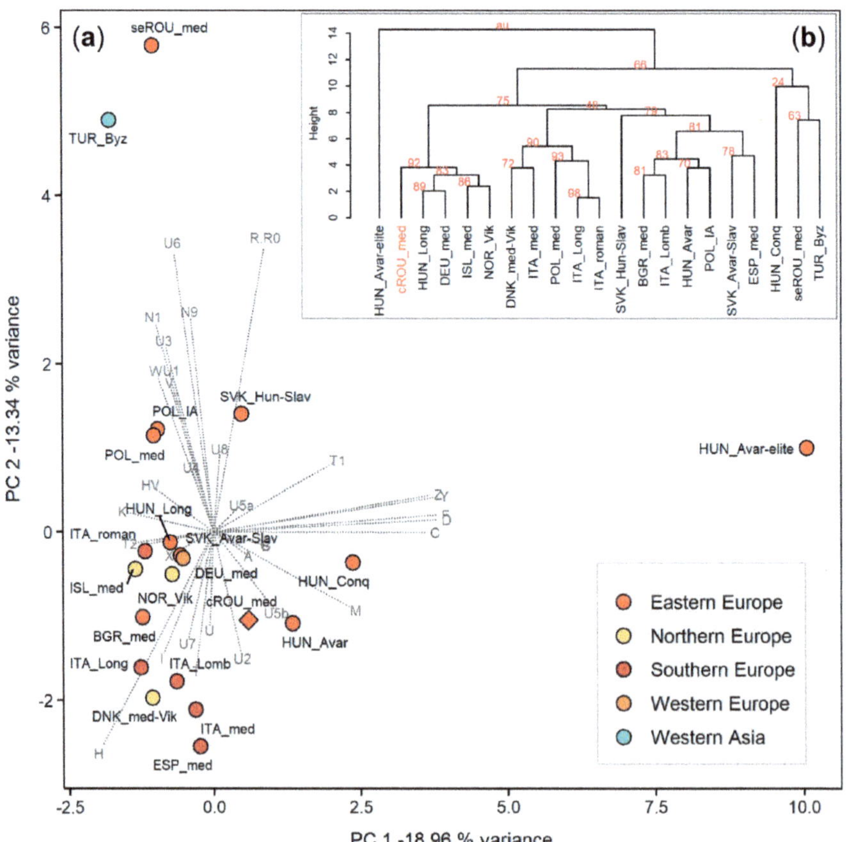

Figure 2. Haplogroup-based analyses of 21 historical populations. (**a**) PCA plot of the first two principal components; (**b**) Ward type hierarchical clustering (AU *p*-values in percent are given as red numbers on the dendogram). The investigated population from Feldioara (cROU_med) is clustered with early medieval individuals from Bavaria (DEU_med), Longobards from Hungary (HUN_Long), Vikings from Norway (NOR_Vik), and Early medieval Icelanders (ISL_med). This cluster is distinct from the one formed by mid-Byzantine population from southwestern Anatolia (TUR_Byz) and a medieval population from southeastern Romania (seROU_med). Abbreviations and mtDNA haplogroup frequencies are reported in Table S4.

Pairwise genetic distances were calculated based on HVR-I sequences of uniform length using the same dataset as for the PCA. The linearized Slatkin Fst values and significant *p* values were visualized on a level plot (Figure 3). The studied population showed non-significant differences from the majority of historical populations, but not from the medieval population from a Hungarian–Slavic Contact zone, the medieval Italians, or the Longobards (also from Italy). The medieval population from Feldioara showed the lowest genetic distances from eight ancient populations of which four were placed under the same branch in the hierarchical clustering diagram (HUN_Long, DEU_med, ISL_med, and Nor_Vik). The other populations showing low genetic distances are: seROU_med, TUR_Byz, POL_med, and POL_IA. The exact Fst values and their corresponding p-values are listed in Table S5. Considering the small sample size analyzed in this study, the low resolution of mitochondrial control region data, and the unequal population datasets used for comparisons, these Fst results might not reflect the real genetic relationships between the compared historical populations, and as such no further interpretations and conclusions can be drawn from this data.

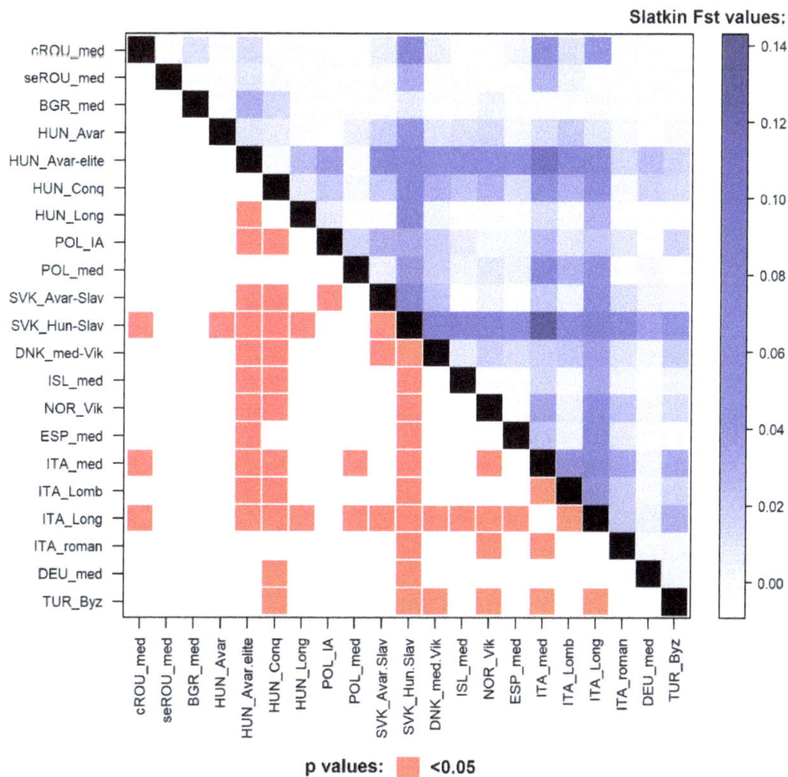

Figure 3. Levelplot of the linearized Slatkin population differentiation (Fst) values and significant p-values. Lower left corner: significant p-values (<0.05) are indicated in red. Upper right corner: larger Slatkin Fst values indicating greater genetic distances are marked by dark blue shades. The exact Fst and p-values are given in Table S5.

The PCA plot based on mtDNA haplogroup frequencies of modern Eurasian populations shows a fair separation of Asian and European populations along the first two components (explaining 40.66% variance) (Figure 4). The present-day European populations are more closely grouped in comparison to the Asian populations which are more dispersed. Apart from the populations originating in Northern Europe which seem to be more densely associated, no other apparent geographical grouping can be identified among European populations. The analyzed population from Transylvania is located on the verge of the European populations' cluster on PC2 and has a higher genetic affinity to some Slavic populations (Ukraine and Slovakia) and modern populations of the Baltic region (Finland, Latvia, and Estonia) than to the others. Most probably, the association of the Feldioara population to the Northern European ones, particularly to modern Finns, is promoted by the relatively high frequency of U lineages in the maternal genetic composition of the investigated group, also present in notable proportions in northern Fenno-Scandinavia and the East Baltic region [66].

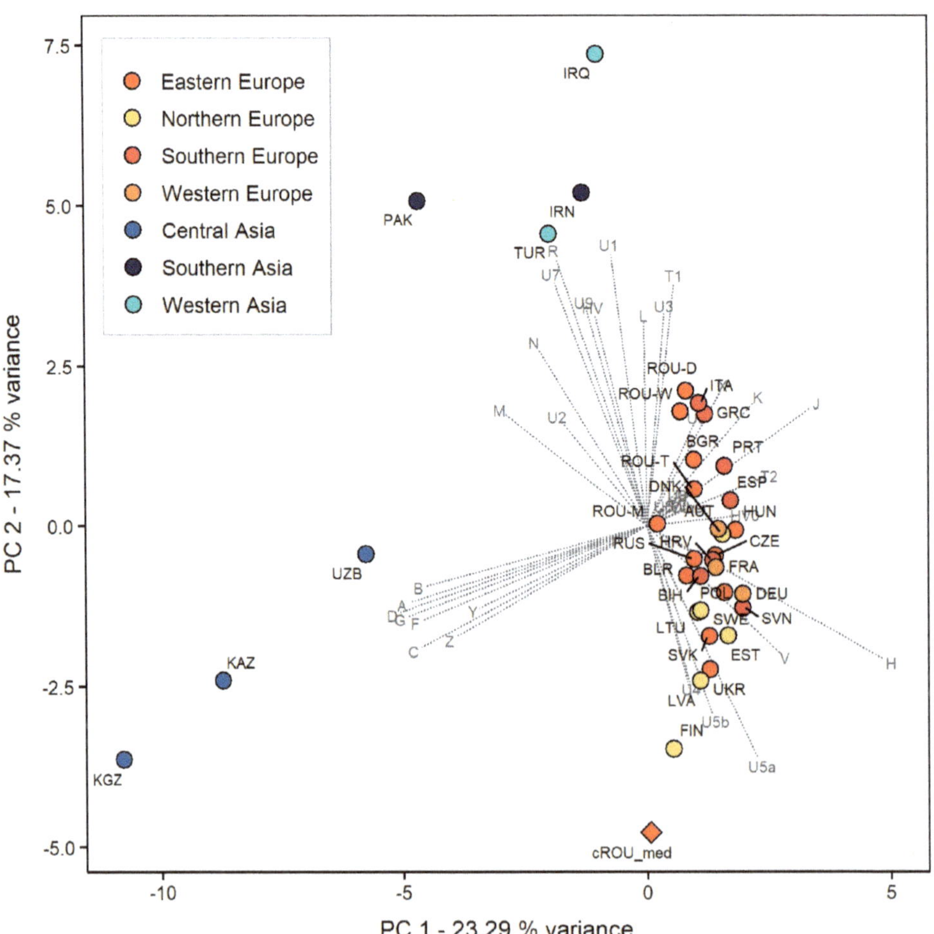

Figure 4. PCA plot of the investigated population (cROU_med) and modern Eurasian populations, based on mtDNA frequencies. MtDNA data for modern Eurasian populations were retrieved from previously published data in Rusu et al. 2018 [20] and references therein.

The medieval Feldioara population showed significant differences ($p < 0.05$) from two modern populations located in Central Asia (Kazakhstan and Kyrgyzstan), while the lowest genetic distances were observed relative to contemporary Slavic populations, mainly from Eastern Europe (BLR, SVK, UKR, LTU, and RUS). The calculated Fst values between pairs of populations and their corresponding p-values, along with reference population information are given in Table S6. The output of the genetic distance analysis was plotted by a MDS (Figure 5). The distribution pattern of modern Eurasian populations reflected by the MDS plot resembles the image of the PCA as most of the European populations are closely grouped, whereas the Asian ones are more scattered along the first two coordinates. The cROU_med is placed in the proximity of modern Romanians from Transylvania and of its two adjacent historical regions of Romania (Wallachia and Moldavia). In contrast, the present-day population from Dobruja, the most geographically distant territory from Transylvania, is less genetically connected to the population investigated in this study. However, the genetic distances between the investigated population and modern Eurasian populations do not allow for unequivocal conclusions as the resulting configuration is sketched only by a partial marker in the genome and is influenced by the low sample size of Feldioara group.

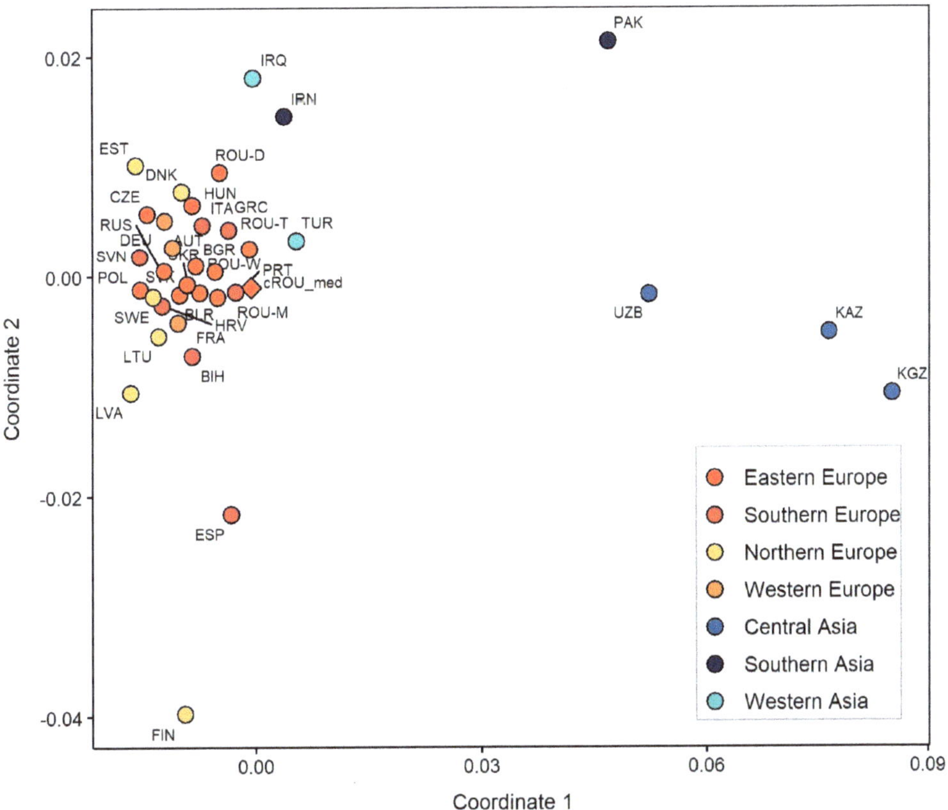

Figure 5. MDS plot of the investigated population (cROU_med) and modern Eurasian populations. The MDS plot was performed using linearized Slatkin Fst values, detailed in Table S6. Stress value is 0.1085145.

4. Conclusions

To gain insights into the genetic relatedness of a Middle Age population of central Romania, we have successfully investigated the mtDNA of 13 individuals excavated from the 12th–13th century necropolis of Feldioara. All 13 individuals analyzed in this study display a different mitochondrial lineage, reflecting a lack of close maternal relationships, even between individuals interred closely by one another. This kind of diversity may characterize a relatively recently established group, with high input of individuals with different geographic origins. West Eurasian haplogroups (H, HV, J, I, U4, and U5) dominate among the sampled individuals, the exception being haplogroup C. Based on available published data, the C4a2 variant identified in one Feldioara sample, can be ultimately traced to the Eastern Asian Bronze Age. A possible scenario of the influx of this lineage into Feldioara might be related to the arrival of Hungarian conquerors in the Carpathian Basin at the end of the 9th century. The population genetic comparisons performed in this study potentially reveal some connections to other historical or modern Eurasian population, but these relations should be treated with care due to the low resolution of the emerging pattern shaped by the low samples size and the use of a partial mtDNA genomic marker. A much more complex and accurate picture of ancestry for the Transylvanian region might be provided by future genome-wide studies and increased number of samples.

Supplementary Materials: The following are available online at https://www.mdpi.com/2073-4425/12/3/436/s1, Figure S1: Overview of graves distribution within Feldioara necropolis (simplified from [14]), Table S1: MtDNA profile of the researchers, Table S2: Samples—processing details, Table S3: MtDNA profile of the investigated samples, Table S4: MtDNA haplogroup frequencies used for PCA with 21 historical populations, Table S5: Fst values, *p*-values and Slatkin Fst matrix of the 21 historical populations, Table S6: Fst values, *p*-values and Slatkin Fst matrix of the medieval population from Feldioara and modern Eurasian populations.

Author Contributions: Conceptualization, I.R. and B.K.; methodology, I.R., B.K., and A.G.; validation, A.G. and C.M.; formal analysis, I.R. and A.G.; investigation, A.G., C.M., and A.I.; resources, H.L.B., I.R., and A.I.; data curation, A.G. and I.R.; writing—original draft preparation, I.R. and A.G.; writing—review and editing, I.R., H.L.B., A.I., C.M., and B.K.; visualization, I.R.; supervision, H.L.B. and B.K.; project administration, I.R. and B.K.; funding acquisition, I.R. and H.L.B. All authors have read and agreed to the published version of the manuscript.

Funding: This research was funded by Babeș-Bolyai University, grant number GTC 31374; H.L.B was supported by a grant of Ministry of Research and Innovation, CNCS-UEFISCDI, project number PN-III-P4-ID-PCCF-2016-0016; I.R. was supported by Entrepreneurship for innovation through doctoral and postdoctoral research project, POCU/380/6/13/123886, co-financed by the European Social Fund, through the Operational Program for Human Capital 2014- 2020.

Institutional Review Board Statement: Not applicable.

Informed Consent Statement: Not applicable.

Data Availability Statement: The mtDNA sequences presented in this study were submitted to GenBank (https://www.ncbi.nlm.nih.gov/genbank/) under the accession numbers MW508518-MW508530. All other relevant data are within the paper and its Supplementary Materials.

Conflicts of Interest: The authors declare no conflict of interest.

References

1. Curta, F. *Southeastern Europe in the Middle Ages, 500–1250*; Cambridge University Press: Cambridge, UK, 2006.
2. Cocos, R.; Schipor, S.; Hervella, M.; Cianga, P.; Popescu, R.; Banescu, C.; Constantinescu, M.; Martinescu, A.; Raicu, F. Genetic affinities among the historical provinces of Romania and Central Europe as revealed by an mtDNA analysis. *BMC Genet.* **2017**, *18*, 20. [CrossRef] [PubMed]
3. Turchi, C.; Stanciu, F.; Paselli, G.; Buscemi, L.; Parson, W.; Tagliabracci, A. The mitochondrial DNA makeup of Romanians: A forensic mtDNA control region database and phylogenetic characterization. *Forensic Sci. Int. Genet.* **2016**, *24*, 136–142. [CrossRef] [PubMed]
4. Hervella, M.; Izagirre, N.; Alonso, S.; Ioana, M.; Netea, M.G.; de-la-Rua, C. The Carpathian range represents a weak genetic barrier in South-East Europe. *BMC Genet.* **2014**, *15*, 56. [CrossRef] [PubMed]
5. Göckenjan, H. *Hilfsvölker und Grenzwächter im Mittelalterlichen Ungarn*; Steiner: Wiesbaden, Germany, 1972; Volume 5.
6. Nägler, T. *Die Ansiedlung der Siebenbürger Sachsen*; Kriterion-Verlag: Bucharest, Romania, 1979.
7. Zimmermann, H. *Der Deutsche Orden in Siebenbürgen: Eine Diplomatische Untersuchung*; Böhlau Verlag Köln: Weimar, Fermany, 2011; Volume 26.
8. Bosch, E.; Calafell, F.; Gonzalez-Neira, A.; Flaiz, C.; Mateu, E.; Scheil, H.G.; Huckenbeck, W.; Efremovska, L.; Mikerezi, I.; Xirotiris, N.; et al. Paternal and maternal lineages in the Balkans show a homogeneous landscape over linguistic barriers, except for the isolated Aromuns. *Ann. Hum. Genet.* **2006**, *70*, 459–487. [CrossRef] [PubMed]
9. Brandstatter, A.; Egyed, B.; Zimmermann, B.; Duftner, N.; Padar, Z.; Parson, W. Migration rates and genetic structure of two Hungarian ethnic groups in Transylvania, Romania. *Ann. Hum. Genet.* **2007**, *71*, 791–803. [CrossRef] [PubMed]

10. Mendizabal, I.; Lao, O.; Marigorta, U.M.; Wollstein, A.; Gusmao, L.; Ferak, V.; Ioana, M.; Jordanova, A.; Kaneva, R.; Kouvatsi, A.; et al. Reconstructing the population history of European Romani from genome-wide data. *Curr. Biol. CB* **2012**, *22*, 2342–2349. [CrossRef]
11. Csányi, B.; Bogácsi-Szabó, E.; Tömöry, G.; Czibula, Á.; Priskin, K.; Csõsz, A.; Mende, B.; Langó, P.; Csete, K.; Zsolnai, A.; et al. Y-Chromosome Analysis of Ancient Hungarian and Two Modern Hungarian-Speaking Populations from the Carpathian Basin. *Ann. Hum. Genet.* **2008**, *72*, 519–534. [CrossRef]
12. Ádám, V.; Bánfai, Z.; Maász, A.; Sümegi, K.; Miseta, A.; Melegh, B. Investigating the genetic characteristics of the Csangos, a traditionally Hungarian speaking ethnic group residing in Romania. *J. Hum. Genet.* **2020**, *65*, 1093–1103. [CrossRef]
13. Barbarii, L.; Constantinescu, C.; Rolf, B. A study on Y-STR haplotypes in the Saxon population from Transylvania (Siebenburger Sachsen): Is there an evidence for a German origin? *Rom. J. Leg. Med.* **2004**, *12*, 247.
14. Ioniță, A.; Căpățână, D.; Boroffka, N.G.O.; Boroffka, R.; Popescu, A. *Feldioara-Marienburg: Contribuții Arheologice la istoria Țării Bârsei: Archaölogische Beiträge zur Geschichte des Burzenlandes*; Editura Academiei Române: Bucharest, Romania, 2004.
15. Muja, C.; Ioniță, A. Sexual dimorphism and general activity levels as revealed by the diaphyseal external shape and historical evidence: Case study on a medieval population from Transylvania. *Dacia N.S.* **2015**, *59*, 319–327.
16. Ioniță, A. Die Besiedlung des Burzenlandes im 12.-13. Jahrhundert im Lichte der Archäologie. In *Generalprobe Burzenland*; Siebenbürgisches Archiv. Archiv des Vereins für Siebenbürgische Landeskunde; Böhlau Verlag: Vienna, Austria, 2013; Volume Band 42, pp. 107–124.
17. REMPEL, H. Reihengräberfriedhöfe des 8. bis 11. Jahrhunderts aus Sachsen-Anhalt, Sachsen und Thüringen. Deutsche Akademie der Wissenschaften zu Berlin. *Schr. Der Sekt. Für Vor-U. Frühgeschichte* **1966**, *20*, 44–51.
18. Gáll, E. The date of the appearance of the S-ended lock-rings in the Transylvanian Basin. *Ephemer. Napoc.* **2009**, *19*, 157–176.
19. Giesler, J. Untersuchungen zur chronologie der Bijelo Brdo-Kultur. Ein beitrag zur archäologie des 10. und 11. jahrhunderts im Karpatenbecken. *Prähistorische Z.* **1981**, *56*, 3–221. [CrossRef]
20. Rusu, I.; Modi, A.; Vai, S.; Pilli, E.; Mircea, C.; Radu, C.; Urduzia, C.; Pinter, Z.K.; Bodolica, V.; Dobrinescu, C.; et al. Maternal DNA lineages at the gate of Europe in the 10th century AD. *PLoS ONE* **2018**, *13*, e0193578. [CrossRef]
21. Dabney, J.; Knapp, M.; Glocke, I.; Gansauge, M.T.; Weihmann, A.; Nickel, B.; Valdiosera, C.; Garcia, N.; Paabo, S.; Arsuaga, J.L.; et al. Complete mitochondrial genome sequence of a Middle Pleistocene cave bear reconstructed from ultrashort DNA fragments. *Proc. Natl. Acad. Sci. USA* **2013**, *110*, 15758–15763. [CrossRef]
22. Gabriel, M.N.; Huffine, E.F.; Ryan, J.H.; Holland, M.M.; Parsons, T.J. Improved MtDNA sequence analysis of forensic remains using a "mini-primer set" amplification strategy. *J. Forensic Sci.* **2001**, *46*, 247–253. [CrossRef]
23. Tömöry, G.; Csányi, B.; Bogácsi-Szabó, E.; Kalmár, T.; Czibula, A.; Csősz, A.; Priskin, K.; Mende, B.; Langó, P.; Downes, C.S.; et al. Comparison of maternal lineage and biogeographic analyses of ancient and modern Hungarian populations. *Am. J. Phys. Anthropol.* **2007**, *134*, 354–368. [CrossRef]
24. Andrews, R.M.; Kubacka, I.; Chinnery, P.F.; Lightowlers, R.N.; Turnbull, D.M.; Howell, N. Reanalysis and revision of the Cambridge reference sequence for human mitochondrial DNA. *Nat. Genet.* **1999**, *23*, 147. [CrossRef]
25. van Oven, M. PhyloTree Build 17: Growing the human mitochondrial DNA tree. *Forensic Sci. Int. Genet. Suppl. Ser.* **2015**, *5*, e392–e394. [CrossRef]
26. Weissensteiner, H.; Pacher, D.; Kloss-Brandstatter, A.; Forer, L.; Specht, G.; Bandelt, H.J.; Kronenberg, F.; Salas, A.; Schonherr, S. HaploGrep 2: Mitochondrial haplogroup classification in the era of high-throughput sequencing. *Nucleic Acids Res.* **2016**, *44*, W58–W63. [CrossRef]
27. Ottoni, C.; Ricaut, F.X.; Vanderheyden, N.; Brucato, N.; Waelkens, M.; Decorte, R. Mitochondrial analysis of a Byzantine population reveals the differential impact of multiple historical events in South Anatolia. *Eur. J. Hum. Genet. Ejhg* **2011**, *19*, 571–576. [CrossRef]
28. Vai, S.; Ghirotto, S.; Pilli, E.; Tassi, F.; Lari, M.; Rizzi, E.; Matas-Lalueza, L.; Ramirez, O.; Lalueza-Fox, C.; Achilli, A.; et al. Genealogical relationships between early medieval and modern inhabitants of Piedmont. *PLoS ONE* **2015**, *10*, e0116801. [CrossRef] [PubMed]
29. Vai, S.; Brunelli, A.; Modi, A.; Tassi, F.; Vergata, C.; Pilli, E.; Lari, M.; Susca, R.R.; Giostra, C.; Baricco, L.P.; et al. A genetic perspective on Longobard-Era migrations. *Eur. J. Hum. Genet.* **2019**, *27*, 647–656. [CrossRef] [PubMed]
30. Csősz, A.; Szécsényi-Nagy, A.; Csákyová, V.; Langó, P.; Bódis, V.; Köhler, K.; Tömöry, G.; Nagy, M.; Mende, B.G. Maternal Genetic Ancestry and Legacy of 10(th) Century AD Hungarians. *Sci. Rep.* **2016**, *6*, 33446. [CrossRef] [PubMed]
31. Csáky, V.; Gerber, D.; Koncz, I.; Csiky, G.; Mende, B.G.; Szeifert, B.; Egyed, B.; Pamjav, H.; Marcsik, A.; Molnár, E.; et al. Genetic insights into the social organisation of the Avar period elite in the 7th century AD Carpathian Basin. *Sci. Rep.* **2020**, *10*, 948. [CrossRef] [PubMed]
32. Šebest, L.; Baldovič, M.; Frtús, A.; Bognár, C.; Kyselicová, K.; Kádasi, L'.; Beňuš, R. Detection of mitochondrial haplogroups in a small avar-slavic population from the eigth–ninth century AD. *Am. J. Phys. Anthropol.* **2018**, *165*, 536–553. [CrossRef] [PubMed]
33. Krzewińska, M.; Bjornstad, G.; Skoglund, P.; Olason, P.I.; Bill, J.; Götherström, A.; Hagelberg, E. Mitochondrial DNA variation in the Viking age population of Norway. *Philos. Trans. R. Soc. Lond. Ser. B Biol. Sci.* **2015**, *370*, 20130384. [CrossRef] [PubMed]
34. Melchior, L.; Lynnerup, N.; Siegismund, H.R.; Kivisild, T.; Dissing, J. Genetic diversity among ancient Nordic populations. *PLoS ONE* **2010**, *5*, e11898. [CrossRef] [PubMed]

35. Alzualde, A.; Izagirre, N.; Alonso, S.; Alonso, A.; Albarrán, C.; Azkarate, A.; de la Rúa, C. Insights into the "isolation" of the Basques: mtDNA lineages from the historical site of Aldaieta (6th-7th centuries AD). *Am. J. Phys. Anthropol.* **2006**, *130*, 394–404. [CrossRef]
36. Guimaraes, S.; Ghirotto, S.; Benazzo, A.; Milani, L.; Lari, M.; Pilli, E.; Pecchioli, E.; Mallegni, F.; Lippi, B.; Bertoldi, F.; et al. Genealogical discontinuities among Etruscan, Medieval, and contemporary Tuscans. *Mol. Biol. Evol.* **2009**, *26*, 2157–2166. [CrossRef]
37. Nesheva, D.V.; Karachanak-Yankova, S.; Lari, M.; Yordanov, Y.; Galabov, A.; Caramelli, D.; Toncheva, D. Mitochondrial DNA Suggests a Western Eurasian Origin for Ancient (Proto-) Bulgarians. *Hum. Biol.* **2015**, *87*, 19–28. [CrossRef]
38. Neparaczki, E.; Maroti, Z.; Kalmar, T.; Kocsy, K.; Maar, K.; Bihari, P.; Nagy, I.; Fothi, E.; Pap, I.; Kustar, A.; et al. Mitogenomic data indicate admixture components of Central-Inner Asian and Srubnaya origin in the conquering Hungarians. *PLoS ONE* **2018**, *13*, e0205920. [CrossRef]
39. Juras, A.; Dabert, M.; Kushniarevich, A.; Malmström, H.; Raghavan, M.; Kosicki, J.Z.; Metspalu, E.; Willerslev, E.; Piontek, J. Ancient DNA reveals matrilineal continuity in present-day Poland over the last two millennia. *PLoS ONE* **2014**, *9*, e110839. [CrossRef] [PubMed]
40. Csákyová, V.; Szécsényi-Nagy, A.; Csősz, A.; Nagy, M.; Fusek, G.; Langó, P.; Bauer, M.; Mende, B.G.; Makovický, P.; Bauerová, M. Maternal Genetic Composition of a Medieval Population from a Hungarian-Slavic Contact Zone in Central Europe. *PLoS ONE* **2016**, *11*, e0151206. [CrossRef] [PubMed]
41. Helgason, A.; Lalueza-Fox, C.; Ghosh, S.; Sigurethardóttir, S.; Sampietro, M.L.; Gigli, E.; Baker, A.; Bertranpetit, J.; Arnadóttir, L.; Thornorsteinsdottir, U.; et al. Sequences from first settlers reveal rapid evolution in Icelandic mtDNA pool. *Plos Genet.* **2009**, *5*, e1000343. [CrossRef] [PubMed]
42. Rusu, I.; Modi, A.; Radu, C.; Mircea, C.; Vulpoi, A.; Dobrinescu, C.; Bodolică, V.; Potârniche, T.; Popescu, O.; Caramelli, D.; et al. Mitochondrial ancestry of medieval individuals carelessly interred in a multiple burial from southeastern Romania. *Sci. Rep.* **2019**, *9*, 961. [CrossRef] [PubMed]
43. Veeramah, K.R.; Rott, A.; Groß, M.; van Dorp, L.; López, S.; Kirsanow, K.; Sell, C.; Blöcher, J.; Wegmann, D.; Link, V.; et al. Population genomic analysis of elongated skulls reveals extensive female-biased immigration in Early Medieval Bavaria. *Proc. Natl. Acad. Sci. USA* **2018**, *115*, 3494. [CrossRef] [PubMed]
44. Stolarek, I.; Handschuh, L.; Juras, A.; Nowaczewska, W.; Kóčka-Krenz, H.; Michalowski, A.; Piontek, J.; Kozlowski, P.; Figlerowicz, M. Goth migration induced changes in the matrilineal genetic structure of the central-east European population. *Sci. Rep.* **2019**, *9*, 6737. [CrossRef]
45. Emery, M.V.; Duggan, A.T.; Murchie, T.J.; Stark, R.J.; Klunk, J.; Hider, J.; Eaton, K.; Karpinski, E.; Schwarcz, H.P.; Poinar, H.N.; et al. Ancient Roman mitochondrial genomes and isotopes reveal relationships and geographic origins at the local and pan-Mediterranean scales. *J. Archaeol. Sci. Rep.* **2018**, *20*, 200–209. [CrossRef]
46. R Core, T. *R: A Language and Environment for Statistical Computing*; R Foundation for Statistical Computing: Vienna, Austria, 2020.
47. Ward, J.H. Hierarchical grouping to optimize an objective function. *J. Am. Stat. Assoc.* **1963**, *58*, 236–244. [CrossRef]
48. Suzuki, R.; Terada, Y.; Shimodaira, H. *pvclust: Hierarchical Clustering with P-Values via Multiscale Bootstrap Resampling*, R package version 2.2-0; 2019. Available online: https://CRAN.R-project.org/package=pvclust (accessed on 25 January 2021).
49. Excoffier, L.; Lischer, H.E. Arlequin suite ver 3.5: A new series of programs to perform population genetics analyses under Linux and Windows. *Mol. Ecol. Resour.* **2010**, *10*, 564–567. [CrossRef] [PubMed]
50. Darriba, D.; Posada, D.; Kozlov, A.M.; Stamatakis, A.; Morel, B.; Flouri, T. ModelTest-NG: A New and Scalable Tool for the Selection of DNA and Protein Evolutionary Models. *Mol. Biol. Evol.* **2019**, *37*, 291–294. [CrossRef] [PubMed]
51. Tamura, K.; Nei, M. Estimation of the number of nucleotide substitutions in the control region of mitochondrial DNA in humans and chimpanzees. *Mol. Biol. Evol.* **1993**, *10*, 512–526. [CrossRef] [PubMed]
52. Oksanen, J.; Blanchet, F.G.; Friendly, M.; Kindt, R.; Legendre, P.; McGlinn, D.; Minchin, P.R.; O'Hara, R.B.; Simpson, G.L.; Solymos, P.; et al. *Vegan: Community Ecology Package*, R package version 2.5-6; 2019. Available online: https://CRAN.R-project.org/package=vegan (accessed on 25 January 2021).
53. Nikitin, A.G.; Newton, J.R.; Potekhina, I.D. Mitochondrial haplogroup C in ancient mitochondrial DNA from Ukraine extends the presence of East Eurasian genetic lineages in Neolithic Central and Eastern Europe. *J. Hum. Genet.* **2012**, *57*, 610–612. [CrossRef]
54. Duggan, A.T.; Whitten, M.; Wiebe, V.; Crawford, M.; Butthof, A.; Spitsyn, V.; Makarov, S.; Novgorodov, I.; Osakovsky, V.; Pakendorf, B. Investigating the Prehistory of Tungusic Peoples of Siberia and the Amur-Ussuri Region with Complete mtDNA Genome Sequences and Y-chromosomal Markers. *PLoS ONE* **2013**, *8*, e83570. [CrossRef] [PubMed]
55. Pilipenko, A.S.; Trapezov, R.O.; Cherdantsev, S.V.; Babenko, V.N.; Nesterova, M.S.; Pozdnyakov, D.V.; Molodin, V.I.; Polosmak, N.V. Maternal genetic features of the Iron Age Tagar population from Southern Siberia (1st millennium BC). *PLoS ONE* **2018**, *13*, e0204062. [CrossRef]
56. de Barros Damgaard, P.; Martiniano, R.; Kamm, J.; Moreno-Mayar, J.V.; Kroonen, G.; Peyrot, M.; Barjamovic, G.; Rasmussen, S.; Zacho, C.; Baimukhanov, N.; et al. The first horse herders and the impact of early Bronze Age steppe expansions into Asia. *Science* **2018**, *360*, eaar7711. [CrossRef]
57. Nikitin, A.G.; Ivanova, S.; Kiosak, D.; Badgerow, J.; Pashnick, J. Subdivisions of haplogroups U and C encompass mitochondrial DNA lineages of Eneolithic–Early Bronze Age Kurgan populations of western North Pontic steppe. *J. Hum. Genet.* **2017**, *62*, 605–613. [CrossRef]

58. Csáky, V.; Gerber, D.; Szeifert, B.; Egyed, B.; Stégmár, B.; Botalov, S.G.e.; Grudochko, I.V.e.; Matveeva, N.P.; Zelenkov, A.S.; Sleptsova, A.V.; et al. Early medieval genetic data from Ural region evaluated in the light of archaeological evidence of ancient Hungarians. *Sci. Rep.* **2020**, *10*, 19137. [CrossRef]
59. Damgaard, P.d.B.; Marchi, N.; Rasmussen, S.; Peyrot, M.; Renaud, G.; Korneliussen, T.; Moreno-Mayar, J.V.; Pedersen, M.W.; Goldberg, A.; Usmanova, E.; et al. 137 ancient human genomes from across the Eurasian steppes. *Nature* **2018**, *557*, 369–374. [CrossRef] [PubMed]
60. Jeong, C.; Wang, K.; Wilkin, S.; Taylor, W.T.T.; Miller, B.K.; Bemmann, J.H.; Stahl, R.; Chiovelli, C.; Knolle, F.; Ulziibayar, S.; et al. A Dynamic 6,000-Year Genetic History of Eurasia's Eastern Steppe. *Cell* **2020**, *183*, 890–904.e829. [CrossRef]
61. Antonio, M.L.; Gao, Z.; Moots, H.M.; Lucci, M.; Candilio, F.; Sawyer, S.; Oberreiter, V.; Calderon, D.; Devitofranceschi, K.; Aikens, R.C.; et al. Ancient Rome: A genetic crossroads of Europe and the Mediterranean. *Science* **2019**, *366*, 708. [CrossRef] [PubMed]
62. Klunk, J.; Duggan, A.T.; Redfern, R.; Gamble, J.; Boldsen, J.L.; Golding, G.B.; Walter, B.S.; Eaton, K.; Stangroom, J.; Rouillard, J.-M.; et al. Genetic resiliency and the Black Death: No apparent loss of mitogenomic diversity due to the Black Death in medieval London and Denmark. *Am. J. Phys. Anthropol.* **2019**, *169*, 240–252. [CrossRef] [PubMed]
63. Olalde, I.; Brace, S.; Allentoft, M.E.; Armit, I.; Kristiansen, K.; Booth, T.; Rohland, N.; Mallick, S.; Szecsenyi-Nagy, A.; Mittnik, A.; et al. The Beaker phenomenon and the genomic transformation of northwest Europe. *Nature* **2018**, *555*, 190–196. [CrossRef] [PubMed]
64. Margaryan, A.; Lawson, D.J.; Sikora, M.; Racimo, F.; Rasmussen, S.; Moltke, I.; Cassidy, L.M.; Jørsboe, E.; Ingason, A.; Pedersen, M.W.; et al. Population genomics of the Viking world. *Nature* **2020**, *585*, 390–396. [CrossRef]
65. Mathieson, I.; Lazaridis, I.; Rohland, N.; Mallick, S.; Patterson, N.; Roodenberg, S.A.; Harney, E.; Stewardson, K.; Fernandes, D.; Novak, M.; et al. Genome-wide patterns of selection in 230 ancient Eurasians. *Nature* **2015**, *528*, 499–503. [CrossRef]
66. Malyarchuk, B.; Derenko, M.; Grzybowski, T.; Perkova, M.; Rogalla, U.; Vanecek, T.; Tsybovsky, I. The peopling of Europe from the mitochondrial haplogroup U5 perspective. *PLoS ONE* **2010**, *5*, e10285. [CrossRef] [PubMed]

Article

Maternal Lineages from 10–11th Century Commoner Cemeteries of the Carpathian Basin

Kitti Maár [1], Gergely I. B. Varga [2], Bence Kovács [2], Oszkár Schütz [1], Zoltán Maróti [2,3], Tibor Kalmár [3], Emil Nyerki [2,3], István Nagy [4,5], Dóra Latinovics [4], Balázs Tihanyi [2,6], Antónia Marcsik [6], György Pálfi [6], Zsolt Bernert [7], Zsolt Gallina [8,9], Sándor Varga [10], László Költő [11], István Raskó [12], Tibor Török [1,2,*,†] and Endre Neparáczki [1,2,†]

1 Department of Genetics, University of Szeged, H-6726 Szeged, Hungary; kitti.maar@gmail.com (K.M.); schutzoszi@gmail.com (O.S.); endre.neparaczki@bio.u-szeged.hu (E.N.)
2 Department of Archaeogenetics, Institute of Hungarian Research, H-1014 Budapest, Hungary; varga.gergely@mki.gov.hu (G.I.B.V.); kovacs.bence.gabor@mki.gov.hu (B.K.); zmaroti@gmail.com (Z.M.); nyerki.emil@mki.gov.hu (E.N.); tihanyi.balazs@mki.gov.hu (B.T.)
3 Department of Pediatrics and Pediatric Health Center, University of Szeged, H-6725 Szeged, Hungary; kalmar.tibor@med.u-szeged.hu
4 SeqOmics Biotechnology Ltd., H-6782 Mórahalom, Hungary; nagyi@seqomics.hu (I.N.); latinovicsd@seqomics.hu (D.L.)
5 Institute of Biochemistry, Biological Research Centre, H-6726 Szeged, Hungary
6 Department of Biological Anthropology, University of Szeged, H-6726 Szeged, Hungary; antonia.marcsik@gmail.com (A.M.); palfigy@bio.u-szeged.hu (G.P.)
7 Department of Anthropology, Hungarian Natural History Museum, H-1083 Budapest, Hungary; bernert.zsolt@nhmus.hu
8 Ásatárs Ltd., H-6000 Kecskemét, Hungary; gallinazsolt@gmail.com
9 Department of Archaeology, Institute of Hungarian Research, H-1014 Budapest, Hungary
10 Ferenc Móra Museum, H-6720 Szeged, Hungary; varga.sandor.arch@gmail.com
11 Rippl-Rónai Municipal Museum with Country Scope, H-7400 Kaposvár, Hungary; koltolaszlo48@gmail.com
12 Institute of Genetics, Biological Research Centre, H-6726 Szeged, Hungary; rasko@brc.hu
* Correspondence: torokt@bio.u-szeged.hu; Tel.: +36-62544104
† These authors jointly supervised this work.

Citation: Maár, K.; Varga, G.I.B.; Kovács, B.; Schütz, O.; Maróti, Z.; Kalmár, T.; Nyerki, E.; Nagy, I.; Latinovics, D.; Tihanyi, B.; et al. Maternal Lineages from 10–11th Century Commoner Cemeteries of the Carpathian Basin. *Genes* **2021**, *12*, 460. https://doi.org/10.3390/genes12030460

Academic Editors: Jennifer A. Leonard and David Caramelli

Received: 26 January 2021
Accepted: 16 March 2021
Published: 23 March 2021

Abstract: Nomadic groups of conquering Hungarians played a predominant role in Hungarian prehistory, but genetic data are available only from the immigrant elite strata. Most of the 10–11th century remains in the Carpathian Basin belong to common people, whose origin and relation to the immigrant elite have been widely debated. Mitogenome sequences were obtained from 202 individuals with next generation sequencing combined with hybridization capture. Median joining networks were used for phylogenetic analysis. The commoner population was compared to 87 ancient Eurasian populations with sequence-based (Fst) and haplogroup-based population genetic methods. The haplogroup composition of the commoner population markedly differs from that of the elite, and, in contrast to the elite, commoners cluster with European populations. Alongside this, detectable sub-haplogroup sharing indicates admixture between the elite and the commoners. The majority of the 10–11th century commoners most likely represent local populations of the Carpathian Basin, which admixed with the eastern immigrant groups (which included conquering Hungarians).

Keywords: ancient mitogenome; Hungarian commoners; Carpathian Basin

1. Introduction

Hungarian history was profoundly determined by the conquering Hungarians (succinctly, the Conquerors), who arrived at the Carpathian Basin from the Eastern European steppe at the end of the 9th century AD as an alliance of seven tribes. The leaders of the alliance (Álmos and his son Árpád) founded a steppe state upon the ashes of the Avar Khaganate [1,2], and their descendants later established the Hungarian Kingdom.

The archaeological legacy of the Conquerors is well defined, especially in the small 10th century cemeteries of the military leader strata whose grave finds included precious metal jewels and costume ornaments as well as decorated horse riding- and weapon-related grave goods [3]. Most of the larger cemeteries attributed to the common people are dated somewhat later, to the 10–12th centuries. People in these so-called village cemeteries were buried with simpler jewels and grave goods, with the sporadic appearance of weapons or harness accessories. There is a general agreement that elite graves with typical grave goods represent first- or second-generation immigrant Conquerors, but the affiliations of people in the village cemeteries are far less clear. For 50 years, they were identified with the Bijelo Brdo culture of the local Slavic people, until their relation to the Conquerors was recognized in 1962 [4] (see Appendix A for details), but to what extent they can be identified with the immigrants as opposed to the previous local population is not yet clear. The answer to this question considerably determines the historical interpretation of the conquest and subsequent events in the Carpathian Basin, and genetic data may contribute to clarifying this issue.

Hitherto, most genetic studies were focused on the elite graves, as these promised an answer for the origin of the immigrant groups. In [5], 76 individuals were selected from 23 cemeteries mainly representing the 10th century elite, and 23% of the maternal lineages identified from hypervariable region (HVR) sequences were east Eurasian and 77% were west Eurasian. Another study, [6], aimed at characterizing the population of entire group of elite cemeteries, sequencing 102 mitogenomes (30% of which had Central–Inner Asian maternal ancestry, while most of the remaining lineages originated from western Eurasia). Y-chromosome studies [7] found that male lineages had similar phylogeographic compositions to female ones. Thus, all studies had congruent results, inferring that the Conqueror elite population originated from an admixture of Asian and European groups on the Pontic steppe.

This raises the question of whether the commoners were genetically similar to the elite, and, if so, could they be one and the same population, or did the poorer strata have a different origin? This question was addressed in the first HVR-based study [8], in which 27 selected graves from 15 cemeteries were grouped according to the type of grave goods present, and the population with "classical" grave goods were found to contain a higher proportion of east Eurasian haplogroups (Hgs) than the group with poor archaeological remains. However, this conclusion was based on a small sample size and a low resolution HVR study, and a systematic characterization of the commoner population with a representative dataset has not been performed yet.

We set out to implement a comprehensive study in this matter, and to this end, we selected eight cemeteries archaeologically evaluated as belonging to the 10–11th century commoners, from which we obtained 202 whole mitogenome sequences. Phylogenetic analysis was performed to illuminate the origin of each maternal sub-Hg of the studied remains. We compared the mitochondrial haplogroup compositions of the commoner and elite populations to find out their genetic relationship and applied different population genetic methods to elucidate the relationship of the commoners with other ancient Eurasian populations. For this reason, we also built a comprehensive database of ancient Eurasian populations, which included all available published mitogenome data.

2. Materials and Methods

2.1. Archeological Background

In contrast to the small 10th century cemeteries with characteristic grave goods [9] representing the conquering Hungarian elite (ConqE), archaeologists classify large 10–11th century cemeteries containing poor grave goods with the sporadic appearance of ConqE findings (see Appendix A for details) as belonging to Hungarian commoners (ConqC). We collected petrous bones (or where these were unavailable, teeth) from 229 human remains from 10 archeological sites (Figure 1) associated with Hungarian commoners.

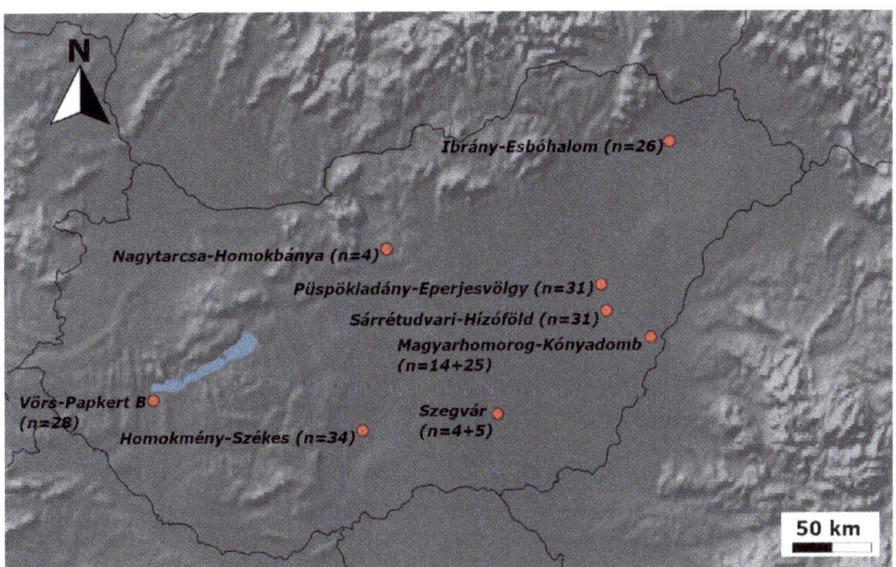

Figure 1. The locations of the graveyards of the Hungarian commoners (ConqC) under study. Sample size is indicated next to cemetery names; two numbers in Magyarhomorog and Szegvár indicate that two nearby cemeteries were sampled. The map was generated using QGIS 3.12.0 [10].

We made an effort to carry out representative sampling. Thus, graves were selected from each section of the cemeteries (including males and females from burials both with and without grave goods and all anthropological types). The number of collected, processed and analyzed samples from each cemetery is summarized in Table 1.

Table 1. Summary of the studied sample size from each cemetery. The mitogenome sequence was obtained after hybridization capture or whole genome sequencing (WGS) as indicated. Samples represent 10–11th century commoners except 14 samples from Magyarhomorog and 20 samples from Vörs-Papkert B. As indicated, we also co-analyzed 13 previously published mitogenomes with new data from this study.

Archaeological Site	Dating (Century CE) Type of Cemetery	No. of Graves	Collected Samples in This Study	Obtained Mitogenomes in This Study (Capture or WGS)	Previously Published Mitogenomes	No. of Samples Analyzed
Sárrétudvari-Hízóföld	10th commoner	262	32	31 (capture)	8	39
Püspökladány-Eperjesvölgy	10–12th commoner	637	36	31 (capture)		31
Ibrány-Esbóhalom	10–11th commoner	269	32	26 (capture)		26
Homokmégy-Székes	10–11th commoner	206	36	34 (capture)		34
Magyarhomorog-Kónyadomb (10–11th century commoner)	10–11th commoner	523	27	25 (capture)	1	26
Magyarhomorog-Kónyadomb (10th century elite)	10th elite	17	14	14 (capture)		14

Table 1. *Cont.*

Archaeological Site	Dating (Century CE) Type of Cemetery	No. of Graves	Collected Samples in This Study	Obtained Mitogenomes in This Study (Capture or WGS)	Previously Published Mitogenomes	No. of Samples Analyzed
Vörs-Papkert B	8–9th	716	Avar period: 9	8 (capture)		8
	9–10th		Carolingian period: 11	11 (capture)		11
	10–11th commoner		Conquest period: 10	9 (capture)		9
Nagytarcsa-Homokbánya	10–11th commoner	21	4	4 (WGS)		4
Szegvár-Oromdűlő	10–11th commoner	372	7	4 (WGS)	2	6
Szegvár-Szőlőkalja	10th commoner	62	11	5 (WGS)		5
Orosháza-Görbicstanya	10th commoner	3			1	1
Szabadkígyós-Pálliget	10th commoner	17			1	1

The largest 10th century commoner cemetery with 262 graves was excavated in Sárrétudvari-Hízóföld [11] (Appendix A.2.6), with a high proportion of graves containing archery equipment and stirrups. We recovered 31 mitogenomes from this site, and a further 8 sequences were added from our previous study [6,12].

Another large commoner cemetery with 637 graves is located in the nearby Püspökladány-Eperjesvölgy [11] (Appendix A.2.5). This cemetery contains a "pagan" and a "Christian" section. Both sections of the graveyard were sampled and we obtained 31 mitogenomes.

The Ibrány-Esbóhalom commoner cemetery with 269 graves also stretches into the Christian era [13] (Appendix A.2.2). We analyzed 32 remains from this site, resulting in the obtainment of 26 mitogenomes.

We studied 36 remains from the Homokmégy-Székes cemetery excavated at the Duna-Tisza Interfluve [14] with 206 graves, which was referred to by the archaeologist as a "typical cemetery of conquest period commoners" (Appendix A.2.1), and obtained 34 mitogenomes.

Among the studied cemeteries, Magyarhomorog-Kónyadomb [15] (Appendix A.2.3) is an exceptional case, as archaeologically it can be divided into two sections: a small section with 17 individuals was used by the 10th century Conqueror elite, while the larger section with 523 graves of 10–11th century commoners raises the question of potential continuity. We sequenced 14 samples from the elite section and 25 samples from the commoner section.

From the Transdanubia region, we included the Vörs-Papkert-B cemetery [16] (Appendix A.2.9), the 716 excavated burials of which are mostly from the late Avar and Carolingian periods. However, 33 people can be dated to the time of the Hungarian conquest. The uninterrupted usage of this graveyard raises the possibility that it might represent the same population in the subsequent periods; thus, we sampled graves from each period as indicated in Table 1.

Finally, we complemented our sample set with a few individuals from the Nagytarcsa-Homokbánya (Appendix A.2.4), Szegvár-Oromdűlő (Appendix A.2.7) and Szegvár-Szőlőkalja (Appendix A.2.8) commoner cemeteries, as listed in Table 1. All of the 13 samples came from poor burials or from graves devoid of archaeological grave goods. For a detailed description of the sites and archaeological findings, see Table S1.

2.2. Library Preparation, Sequencing and Hg Assignment

All pre-PCR laboratory procedures leading to next generation sequencing (NGS) were conducted in the common ancient DNA laboratory of the Department of Archaeogenetics of the Institute of Hungarian Research and Department of Genetics, University of Szeged, Hungary. Details concerning the ancient DNA purification, library preparation, hybridization capture, sequencing and sequence analysis method are given in [12]. We used the double stranded library protocol of [17] with double indexing [18]. All libraries were made from partial uracil-DNA glycosylase (UDG)-treated DNA extracts [19]. We estimated the endogenous human DNA content of each library with low coverage shotgun sequencing (Table S2a). Then, the mitogenomes from samples with similar proportions of human DNA content were pooled and enriched together according to [20]. Captured and amplified libraries were purified on MinElute columns. Quantity and quality measurements were performed with the Qubit fluorometric quantification system and the TapeStation automated electrophoresis system (Agilent). A further 13 mitogenome sequences were obtained from whole genome sequencing, as indicated in Table 1 and Table S2.

The adapters of paired-end reads were trimmed with the Cutadapt software [21] in paired end mode. Read quality was assessed with FastQC [22]. Sequences shorter than 25 nucleotides were removed from this dataset. The resulting analysis-ready reads were mapped to the GRCh37.75 human genome reference sequence that also contains the mtDNA revised Cambridge Reference Sequence (rCRS, NC_012920.1) [23] using the Burrows Wheeler Aligner (BWA) v0.7.9 software [24] with the BWA mem algorithm in paired mode and default parameters. Samtools v1.1 [25] was used for sorting and indexing binary alignment map (BAM) files. PCR duplicates were removed using Picard Tools v 1.113 [26]. Ancient DNA damage patterns were assessed using MapDamage 2.0 [27] and read quality scores were modified with the rescale option to account for post-mortem damage. Mitochondrial genome contamination was estimated using the Schmutzi algorithm [28] (Table S2b). Mitochondrial haplogroup (Hg) determination was performed using HaploGrep v2.1.25 [29] (Table S3a). The biological sex of the individuals was identified according to [30] based on the X/Y ratio of the reads obtained from shotgun sequencing.

The raw nucleotide sequence data of the 202 samples were deposited to the European Nucleotide Archive (http://www.ebi.ac.uk/ena) under the accession number: PRJEB40566.

2.3. Assembling an Ancient Eurasian Mitogenome Database

For the phylogenetic and population genetic analyses, we built a database containing 4191 published ancient Eurasian mitogenomes (Table S4). Sequences were downloaded from the NCBI and the European Nucleotide Archive databases. Where it was necessary, mitogenome sequences were sorted out from whole genomes. This database was then augmented with the 202 new mitogenomes from this study. We ordered the published samples into 88 populations based on time range, archaeological site and context, as well as the classification of the published genome data. In cases when populations were under-represented due to a low sample size, we grouped samples from related cultures like Alans and Saltovo-Mayaki, Medieval samples from Italy, Germany and England, Iberian Chalcolithic and Bronze Age samples, Chalcolithic samples from Iran and Turan, early and late Sarmatians, etc. (Table S4).

2.4. Phylogenetic and Population Genetic Study

A sub-set of the published sequences was of poor quality. We excluded sequences with >5% missing data from the phylogenetic and Fst analysis and used 3844 fasta files of ancient sequences and 11,682 fasta files of modern sequences for building median joining (MJ) networks, as described in [6]. The phylo-geographic origins of the samples were assessed from the geographic origin of the nearest Hgs. We distinguished four regions: east Eurasia, west Eurasia, Eurasia and Caucasus–Middle East (Figure S1).

For population genetic analysis, we merged all 169 ConqC data to a single population (Tables S3c and S4), excluding members of the elite Magyarhomorog cemetery as well as

Avar and Caroling samples from the Vörs-Papkert cemetery (excluded samples are color labeled in Table S2b). These were supplemented with 13 commoner mitogenomes published previously [6], as listed in Table 1. The merged ConqC population was compared to the 88 ancient Eurasian groups from the newly assembled mitogenome database, including the previously published military elite strata of the Conquerors [6,12,31], which was supplemented with the Magyarhomorog elite graveyard data from the present study (Table 1, Tables S3c and S4).

Three independent methods were applied to measure the genetic distances of ConqC from other ancient populations. In the first analysis, we reduced the Hg assignments of all samples to major Hgs, which decreased population data to 34 dimensions, which is sufficient to display the main correlations. Then, major Hg frequencies were calculated and principal component analysis (PCA) was conducted, employing the function "prcomp" in R 3.6.3. [32]. We also compared the major Hg frequencies of the ConqC and ConqE groups separately.

In a second approach, a traditional sequence-based method was implemented, calculating pair-wise population differentiation values (Fst) with Arlequin 3.5.2.2 [33] from whole mtDNA genomes, as described in [6]. Multi-dimensional scaling (MDS) was applied on the matrix of linearized Slatkin Fst values [34], and the values were visualized in the two-dimensional space using the cmdscale function implemented in R 3.6.3 [32].

In a third approach, shared haplogroup distance (SHD) values were measured between the populations according to our previous study [6], which calculates the frequency of identical terminal sub-Hgs (the deepest determined Hg level) between populations, as these testify shared ancestry or past admixture.

3. Results

3.1. Sequencing Results and Assigned Haplogroups

We collected a total of 229 samples from the listed sites, but could not obtain DNA from 13 samples. Another 10 samples were excluded from the analysis due to low mitogenome sequence coverage and 3 further samples were excluded due to high contamination values. Using the NGS method combined with target enrichment, we acquired 189 ancient mitogenome sequences, and a further 13 were obtained from whole genome sequencing; thus, we report 202 new mitogenomes in this paper (Table S3a). We obtained 4.2-3068x mitogenome coverage, and the average coverage was 231x. Schmutzi estimated negligible contamination for most of the 202 samples. Seven samples were indicated to carry significant (15–21%) contamination; nonetheless, Schmutzi could determine the endogenous sequence unambiguously for these samples due to high coverage, enabling a correct Hg assignment. For details of the sequencing data, see Table S2. On the grounds of haplogroup determination by HaploGrep 2.0, the 202 samples belong to 154 sub-Hgs and 187 different haplotypes (Table S3a).

3.2. Kinship Analysis

We examined a possible kinship relation between and within cemeteries. We detected 10 pairs of identical mitochondrial haplotypes within cemeteries and 4 pairs between cemeteries (Table S3b), which indicate a potential direct maternal relationship of these individuals, but this of course is not inherent evidence of close family relations.

3.3. Phylogenetic Analysis

As some of the mitochondrial sub-clades may have specific geographical distribution [35,36], we elucidated the phylogenetic relations of each mitogenome sequence using M–J networks, as shown in Figure S1. The closest sequence matches pointed at a well-defined geographical region in most cases, which is indicated next to the phylogenetic trees and is summarized on Figure 2.

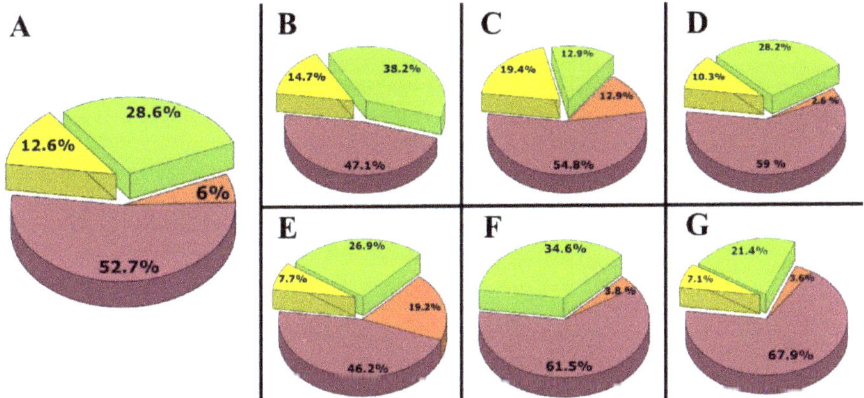

Figure 2. The phylogeographic origin of the ConqC maternal lineages from different cemeteries. Data are summarized from Figure S1 and from a previous study [6]. West Eurasian haplogroups (Hgs) are marked with pink, east Eurasian Hgs are marked with yellow, Eurasian Hgs are marked with green and Caucasus–Middle East Hgs are marked with brown. (**A**) Distribution of the merged data of 182 Hungarian commoner samples from all cemeteries. (**B**–**G**) The phylogeographic distribution of the maternal lineages from individual cemeteries: (**B**) Homokmégy-Székes (n = 34); (**C**) Püspökladány-Eperjesvölgy (n = 31); (**D**) Sárrétudvari-Hízóföld (n = 39); (**E**) Ibrány-Esbóhalom (n = 26); (**F**) Magyarhomorog-Kónyadomb (n = 26, with samples taken just from the commoner part); (**G**) Vörs-Papkert-B (n = 28, including all samples from this cemetery).

Phylogenetic trees revealed that, out of the 182 commoner maternal lineages, 23 were unequivocally derived from east Eurasia and 107 were derived from west Eurasia, while 52 are widespread throughout Eurasia. Out of the western Eurasian lineages, 11 have a primarily Caucasus–Middle East distribution (Figure 2A).

3.4. Haplogroup Composition of Individual Cemeteries

The 34 investigated samples from Homokmégy-Székes belonged to 30 Hgs (Table S3a). As for the lineages, 47.1% were of European origin and 14.7% were of east Eurasian origin, while 38.2% showed general Eurasian distribution (Figure 2B).

From the Püspökladány-Eperjesvölgy cemetery, 31 remains were analyzed. The maternal lineages were classified into 28 Hgs (Table S3a), and they showed 54.8% west Eurasian ancestry, 19.4% east Eurasian ancestry and 12.9% Eurasian ancestry, while 12.9% had a Caucasus–Middle East affinity (Figure 2C).

The newly reported mitogenomes of 31 individuals from Sárrétudvari-Hízóföld belonged to 26 Hgs (Table S3a). In a previous study, the mitochondrial lineage of eight individuals from this cemetery were obtained [6]. Merging these data, 59% of the lineages had west Eurasian ancestry, 10.3% had east Eurasian ancestry, 28.2% had Eurasian ancestry and 2.6% had Caucasian–Middle Eastern maternal ancestry (Figure 2D).

The Ibrány-Esbóhalom cemetery was represented by 26 samples falling to 26 different Hgs (Table S3a). 46.2% of the maternal lineages originated from Europe, 7.7% originated from east Eurasia and 19.2% originated from the Caucasus–Middle East region, while 26.9% of the lineages had a Eurasian distribution (Figure 2E).

We sequenced 14 mitogenomes out of the 17 remains from the elite part of Magyarhomorog-Kónyadomb, and their Hg composition was very similar to those of previously studied elite cemeteries [6]; 35.7% of the lineages were of east Eurasian origin, 42.9% were of European origin and 21.4% were of Eurasian origin (Table S1). The high frequency of N1a1a1a1a and T1a1, as well as the occurence of N1a1a1a1 and D4 in this small cemetery, finds its best parallels in the Karos and Kenézlő elite graveyards [4], supporting the archaeological evaluation; thus, we included these data in the elite dataset (Table S3c).

From the 11–12th century commoner part of Magyarhomorog, we sequenced 25 samples which belonged to 22 mitochondrial Hgs (Table S3a), supplemented with one published sample from this site [6]. From the 26 samples, 61.5% had a west Eurasian origin, 34.6% had an Eurasian origin and 3.8% had a Caucasus–Middle East affinity (Figure 2F); thus, genetic data also corroborated the hypothesis that the large graveyard represents a separable commoner population.

The cemetery of Vörs-Papkert is another special case, as it was used for centuries by successive populations of Avars, Carolingians and Conquerors populations. Evaluating the entire 28 sample set from this cemetery together (Figure 2G) showed a very similar overall picture to that of other commoner cemeteries, with 25 Hgs, 67.9% of which had a west Eurasian origin, 7.1% had an east Eurasian origin, 21.4% had an Eurasian origin and 3.6% had a Caucasus–Middle East affinity. Hg H dominated this graveyard, as 16 out of the 28 remains belonged to Hg H irrespective of historical period. A single D4e4 Hg was detected among the studied ConqC and a single A16 was detected among the Avar period samples as weak signs of Asian impact (Table S1). By all means, for the population genetic analysis, we removed Avar and Carolingian samples from this dataset.

The six ConqC graveyards with a meaningful sample size showed a rather similar overall picture, with an average of 12.6% east Eurasian Hgs almost confined to C and D, which allowed us to infer a similar overall east Eurasian impact throughout cis-Danubia.

We also investigated a few individuals from other commoner cemeteries, namely four samples from Nagytarcsa-Homokbánya, four samples from Szegvár-Oromdűlő and five samples from Szegvár-Szőlőkalja, resulting in two east Eurasian lineages besides the European ones (Table S3a).

We acknowledge that the average of 30 samples per site may poorly represent the individual cemeteries, but the total number of 182 commoner remains (Table S3c) can be regarded as considerably representative for population genetic analysis.

3.5. Population Genetic Analysis

First, we compared the major Hg distribution of the conqueror period elite and commoner populations (Figure 3). The heterogeneity of the major Hg distribution of ConqE is comparable to that of ConqC (22 and 19 main Hgs, respectively); however, the Hg compositions of the two groups show considerable differences. The ratio of east Eurasian major-Hgs in the commoners is 7.69%, contrary to the 19.64% of the elite. The elite contains a broad spectrum of east Eurasian Hgs (A, B, C, D, F, G and Y), while only C and D occur with notable frequency in the commoners, with a single appearance of B.

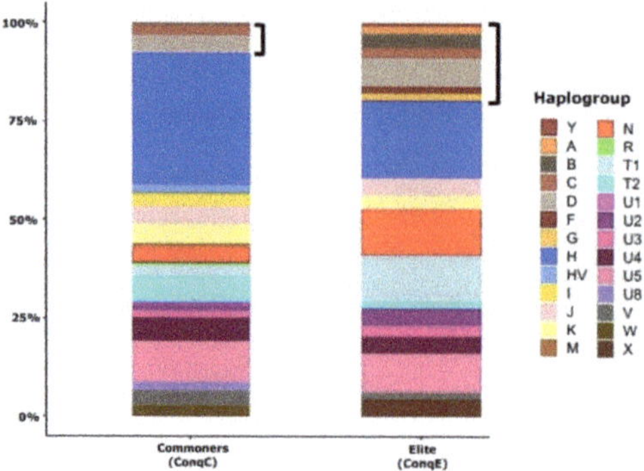

Figure 3. Comparison of the major Hg distributions from ancient Hungarian populations. The major Hg distribution of commoner samples (n = 182) from this study is distinct from that of Conqueror elite samples (n = 112) taken from previous studies [6,12,31], including elite data from Magyarhomorog in the present study. Brackets mark east Eurasian Hgs.

West Eurasian Hgs of ConqC and ConqE also show notable differences: Hgs HV, I, M, R, U1, U8 and W occur with moderate frequencies in commoners, while these are completely absent from the elite population. Three Hgs (N, T1 and X), typically widespread both in east and west Eurasia, show much higher ratio in the elite than in commoners: N's ratios are 11.61% in the elite population and 3.85% in the commoner population; T1's ratios are 11.61% in the elite population and 2.75% in the commoner population; and X's ratios are 4.46% in the elite population and 0.55% in the commoner population. The opposite is true for Hgs H and T2; among commoners, H is the most prevalent Hg with a 33.52% frequency, while in the elite group, its proportion is significantly lower (19.64%); T2 has a 6.59% proportion in the commoner population and a 1.79% proportion in the elite population.

As the Hg composition of the studied commoner samples markedly differs from that of the elite, we measured ConqC's genetic distances from ConqE as well as its distances from 87 published ancient Eurasian populations (Table S5). PCA obtained from the major Hg frequencies of 88 populations (Figure 4) highlights the distance between ConqE and ConqC. The ConqC clustered in the eastern side of the European aggregation, with the closest genetic affinity to Baltic Bronze Age populations, Baltic Iron Age populations, Baltic Medieval populations, Bell Baker Germany and Great Britain Bronze Age populations, and is not far away from the Steppe Early-Middle Bronze Age (Steppe EMBA) population, though these relative distances need to be interpreted with care, as our population dataset certainly incompletely represents the past genetic variability. In contrast, the Conqueror Elite is located between ancient European and Asian populations and its closest clusters are the Sarmatian Iron Age population, the Tien Shan Iron Age population, the Karasuk late Bronze Age population and the two groups suggested to be in connection with the Conquerors [31]: the Cis-Ural Medieval population and the Uyelgi Trans-Ural Medieval population.

Figure 4. The principal component analysis (PCA) plot of the major mtDNA haplogroup distribution (distinguishing Hgs A, B, C, D, F, G, H, HV, I, J, K, L, M, N, N1a, N1b, R, T, T1, T2, U, U1, U2, U3, U4, U5, U6, U7, U8, V, W, X, Y and Z) of 88 Eurasian populations. The abbreviations of the population names are given in Table S4b. Color shadings denote geographic regions as indicated. ConqC and ConqE are highlighted with arrowheads. PC1 separates European populations to the left and Asian populations to the right side. PC2 separates Anatolian–Caucasus groups to the bottom and hunter–gatherers to the top.

In order to further reveal the genetic relationships of ConqC with other ancient groups, we drew an MDS plot (Figure 5) from linearized Slatkin Fst values (Table S5a). Fst distances confirmed that ConqC is nearest to ancient European and Near Eastern populations; in the Pairwise Fst matrix, the closest groups are the European Medieval (0.0098), Anatolia Bronze Age (0.00991), Iceland Medieval (0.01433), pre-Roman (Umbri) Iron Age from Italy (0.01691) and Roman Antiquity (0.01701) groups, followed by other European Bronze Age, Neolithic and Chalcolithic groups. Accordingly these are located close on the MDS plot. On the other hand, Fst data show that ConqC significantly differs from ConqE ($p < 0.00000$); in other words, the probability that the two populations are identical is below 1/100,000.

Figure 5. A multi-dimensional scaling (MDS) plot from the linearized Slatkin Fst values from Table S5a. Abbreviations of the population names are given in Table S4b. European populations are sequestered to the left and Asian populations are sequestered to the bottom right. Color shading denotes geographic regions as indicated. ConqC and ConqE are highlighted with arrowheads.

The novel SHD population genetic method gave similar results, but also revealed new information (Table S5b). ConqE has the smallest SHD distance from ConqC, followed by European populations from the Neolithic to the Medieval periods. It is also notable that the SHD and Fst distances of Steppe EMBA populations are comparable to those of European groups. European Scythian and Scytho–Siberian populations have noteworthy SHD distances as well, indicating that ConqC significantly shared sub-Hgs with these Eurasian steppe populations.

4. Discussion

In this paper, an attempt was made to provide a genetic description of the common people of the Carpathian Basin who lived in the 10–11th centuries during the period of the Hungarian conquest. Of the 202 obtained mitogenomes, 169 belonged to commoners, while 14 samples from the Magyarhomorog cemetery were revealed to represent a small elite graveyard, not related to the adjacent commoner remains. The elite has been shown to comprise around 30% of the east Eurasian Hgs, including characteristic ones like N1a1a1a1a [6] (Table S3c).

The overall Hg composition of the commoner population proved to be significantly different from that of the elite with respect to both east and West Eurasian Hgs, indicating that these two groups likely had different origins. Population genetic analysis definitely clustered ConqC primarily with European and Near Eastern populations, separating them from the elite, suggesting that people with local European origin dominated the ConqC population.

The presence of a non-negligible proportion of east Eurasian Hgs in the ConqC population is a clear sign of admixture with eastern immigrants, presumably with Avars and/or Conquerors. This effect distinguishes ConqC from contemporary European populations, as well as from modern Hungarians, in whom east Eurasian Hgs are negligible. Thus, despite their significant differences, ConqC might have admixed with ConqE to some extent.

This admixture is clearly validated by the SHD method, as ConqC had the smallest SHD distance from ConqE (Table S5b), meaning that out of the studied ancient populations, ConqE shared the highest proportion of identical Hgs with ConqC, best explained by admixture. As the SHD value perfectly represents the common gene pool, the SHD distance of 0.85 indicates a 14% common gene pool between ConqE and ConqC. Out of the 18 shared Hgs, 4 had an east Eurasian origin (Table S3c), so these were very likely transferred from the elite to the commoners. It is especially telling that the most frequent ConqE Hgs (N1a1a1a1, its derivative N1a1a1a1a and T1a1) were present in numerous commoner cemeteries. The east Eurasian N1a1a1a1 ConqE marker most likely originated from the Afanasievo or Sintashta–Tagar cultures [37,38], while, despite its general Eurasian range, a Mongolian Chemurchek–Uyuk–Deer stone–khirigsuur [39] origin of T1a1 in ConqE is very plausible (Table S3c). The close SHD distance of ConqE to Steppe EMBA and Steppe MLBA populations (Table S5b) implies that the Steppe EMBA affinity of ConqC, observed in Figure 4, can also be a consequence of ConqE admixture.

The phylogeographic origin of shared Hgs also signals a possible reciprocal gene flow from ConqC to ConqE, as some of their shared Hgs (H7, K1c1, T2b and V7a) were absent from east Eurasia but had been present in the Carpathian Basin from the Neolithic–Bronze Age, as shown in Table S3c. As a consequence, the 14% common gene pool between ConqE and ConqC cannot be interpreted as a headcount proportion of immigrants and local people. Furthermore, both could have acquired common elements from other unknown populations.

The contemporary local population is descended from previous peoples of the Carpathian Basin, and it has indeed been shown that a large number of people survived to the 10th century from the previous Avar period [40,41]. The Avars also brought along a package of east Eurasian Hgs, and a significant fraction of east Eurasian Hgs which are found in ConqC and are not shared with ConqE (such as B5b4, C4a1b, C5b1a, D4b1, D4e4, D4l2, D4m2a and D5a3a1, as shown in Table S3c). These Hgs are potential candidates for Avar heritage.

5. Conclusions

For more accurate conclusions, future investigations are necessary, including high-resolution genome analysis of commoner and elite cemeteries. Additionally, genome data from the pre-Avar, Avar and later Árpádian populations would provide a more complete picture about the exact contribution of subsequent nomadic migrations to the demographic history of the Carpathian Basin.

Supplementary Materials: The following are available online at https://www.mdpi.com/2073-4425/12/3/460/s1, Figure S1: MJ Networks, Table S1: Summary and archaeological details of studied samples, Table S2a: Shotgun sequence data, Table S2b: Mitogenome sequence data, Table S3a: SNP positions of the mitogenomes, Table S3b: List of identical haplotypes. Table S3c: List of ConqC and ConqE individuals and shared haplogroups Table S4: Ancient mitogenome database, Table S5a: Pairwise Fst matrix of ancient populations, Table S5b: SHD distance matrix of ancient populations.

Author Contributions: Conceptualization, T.T. and E.N. (Endre Neparáczki); data curation, K.M., Z.M., E.N. (Emil Nyerki) and E.N. (Endre Neparáczki); formal analysis, K.M., G.I.B.V., Z.M. and E.N. (Endre Neparáczki); funding acquisition, T.T. and E.N. (Endre Neparáczki); investigation, K.M., B.K., D.L., O.S., B.T., G.I.B.V., I.N. and E.N. (Endre Neparáczki); methodology, T.T., Z.M. and E.N. (Endre Neparáczki); project administration, K.M. and E.N. (Endre Neparáczki); resources, B.T., Z.B., A.M., Z.G., S.V., L.K. and G.P.; software, Z.M.; supervision, T.T. and E.N. (Endre Neparáczki); visualization, K.M. and E.N. (Endre Neparáczki); writing—original draft preparation, K.M. and E.N. (Endre Neparáczki); writing—review & editing: Z.M., I.R., T.K., G.I.B.V. and T.T. All authors have read and agreed to the published version of the manuscript.

Funding: This research was funded by grants from the National Research, Development and Innovation Office (K-124350 to T.T. and TUDFO/5157-1/2019-ITM; TKP2020-NKA-23 to E.N. (Endre Neparáczki)), The House of Árpád Programme (2018–2023) Scientific Subproject: V.1. Anthropological-Genetic portrayal of Hungarians in the Árpadian Age to T.T. and No. VI/1878/2020. certificate number grants to E.N. (Endre Neparáczki). K.M. was supported by ÚNKP-20-3-SZTE-470 New National Excellence Program.

Institutional Review Board Statement: Not applicable.

Informed Consent Statement: Not applicable.

Data Availability Statement: The data presented in this study are openly available in to the European Nucleotide Archive (http://www.ebi.ac.uk/ena) under the accession number: PRJEB40566.

Acknowledgments: We are grateful to our archaeologist colleagues, and especially László Kovács, Eszter Istvánovits, Gábor Lőrinczy, Ibolya M. Nepper and József Szentpéteri for their help. We are also grateful to Miklós Kásler and Gábor Horváth–Lugossy for their encouragement and Szabolcs Tóth for his administrative work.

Conflicts of Interest: I.N. and D.L. at SeqOmics Biotechnology Ltd. and Zs.G. at Ásatárs Ltd. had consulting positions during the time the study was conceived. SeqOmics Biotechnology Ltd. and Ásatárs Ltd. were not directly involved in the design and execution of the experiments or in the writing of the manuscript. This affiliation does not alter our adherence to Genes' policies on sharing data and materials.

Appendix A. Brief Archaeological Background of the Hungarian Commoner Cemeteries

(Detailed archaeological and anthropological description of each graves is provided in Table S1)

Appendix A.1. A Brief Summary of the Problems Associated with the Archeological Categorization of 10–11th Century Cemeteries in the Carpathian Basin

Assessments of the archeological horizon, cemeteries, and individual burials of the 10–11th century Carpathian Basin, which is the period of the Hungarian conquest and state formation, have undergone significant changes over the past 150 years. The first major summary and categorization of the finds was made by József Hampel [3] based on dating patterns (tombs that could be dated with a coin, tombs that could not be dated with a coin, and stray finds). However, this was soon replaced by his new classification based on ethnicity in which two groups were distinguished: the findings of newly arrived conquerors (the time covering ca. 150 years, 895–1050) buried with horse-riding- and weapon-related grave goods, often with coins of foreign origin (Western European, Byzantine, Arabian), were referred to as Group A and the local population resting in cemeteries composed of rows of graves and buried with simpler jewels and grave goods, often with coins of the kings of the House of Árpád, were referred to as Group B. The latter group later spread in the public consciousness as the Bijelo Brdo culture and was defined as having Slavic ethnicity.

In the first half of the 20th century the number of archeological findings of the 10–11th century period increased rapidly. During the collection and systematization of these finds, Béla Szőke noticed that Group A and Group B cannot be sharply separated, because the findings of the two groups often appear simultaneously in the cemeteries and the cemeteries composed of rows of graves are also present in the central areas of the 10th century Hungarian territory. Based on his observations, he believed that the differences did not indicate different ethnicities, rather they reflected the social status. Thus he developed a new kind of systematization in which the ethnic-based classification was replaced by a division into social categories [4]. Hampel's Group A was identified with the leading and middle class of the conquerors, and, in connection with Hampel's Group B, he assumed that they were the findings of the commoner people of the 11th century who were mostly of Hungarian origin but often mixed with other ethnicities (e.g., in the peripheral areas). However, the analysis of the archeological material related to the large-scale excavations of the second half of the 20th century already highlighted the problems of the method and its categories (e.g., [42]); Szőke's system was used until the last decades to classify individual burials and cemeteries. In addition to the fact that international research has shown that it is not possible, or very difficult, to distinguish legal and social categories using archeological methods alone [43,44], the social categories have not been sufficiently defined, so, in the field of practice, cemeteries were classified in a highly subjective way. On the other hand, the size of the cemetery significantly influenced the evaluation of the findings (e.g., the presence of 3–4 graves with horse-riding- and weapon-related grave goods in a cemetery of 10–20 graves already elevated the given cemetery to the middle class; however, the same ratio in a cemetery composed of 100 graves resulted in the site being classified as a commoner cemetery). However, based on the most recent investigations, only a low percentage of the 10th century cemeteries can be considered as fully excavated, often resulting in erroneous conclusions [42]. In addition, the cemeteries frequently contained a mix of "rich" and "poor" grave goods or even burials without known grave goods, which draws attention to the fact that a cemetery may not necessarily correspond to a particular social category.

István Bóna attempted to reinterpret the 10–11th century cemeteries by distinguishing two types of cemeteries [45]. The so-called military cemeteries were characterized by the rather low number of graves covering one or two generations and often showing a male surplus, which he considered were mostly used in the first half of the 10th century, while the larger village cemeteries with 30–100 graves were used later, during the 10–12th centuries.

This criteria system was further developed by László Kovács, who, breaking with the ethnic and social classification, based the assessment of the sites on the chronological characteristics and quantity of the burials [46]. Within the archeological horizon of the 10–12th centuries, he distinguished two main types in connection with the supposed form of the possibly related settlement: the so-called quarter cemetery and the village cemetery. The quarter group included smaller cemeteries with 5/10–50/75 burials, which were used for a shorter time period and were abandoned between the second half of the 10th century and the turn of the millennium. These cemeteries were characterized by mostly the types of objects and burial customs associated with the conquerors, and at the same time the coins of the kings of the Árpád Dynasty were completely missing. The other main group, the village cemeteries, was characterized by a larger number of graves of 50–200 and above, and their use can be traced back to the 10th, 10–11th, and 10–12th centuries, depending on the sub-type. In connection with the grouping, it may be a problem to separate the 10th century quarter type and the 10th century village type, as there are overlaps based on the time of use and the number of graves, so sometimes the richness and quality of the finds may make the difference [42], but problems with the representational value and symbolic aspects of the objects [47,48] may mislead the conclusions. Questions arise relating to the evaluation of cemeteries dating from the 10th century which were in use through the 11–12th centuries, because based on the written sources concerning the period of the foundation of the Hungarian Kingdom and the early Árpádian age, as well as the results of micro-regional archeological research, significant internal population movements and organized settlements must be taken into account. With the help of archeological methods alone, it is not possible to determine beyond any doubt whether the 10th century and later cemetery parts are contiguous with each other or whether a population change has taken place in the meantime [42]. However, this can also significantly affect the grouping of cemeteries (e.g., a site can be assessed either as a larger 10–11th century village cemetery or as separately having a 10th century quarter/village cemetery and a 11th century village cemetery).

Several methods have been developed since the beginning of research to categorize 10–11/12th century cemeteries and burials. Each method has its advantages and disadvantages depending on the issues raised by the particular research. Being aware of all this, during the short archeological summary of the cemeteries involved in our research, we tried to describe their characteristics objectively; however, in each case, we indicated where the given site was classified in the Szőke and Kovács classifications.

Appendix A.2. Archeological and Anthropological Description of the Studied Cemeteries

Appendix A.2.1. Homokmégy–Székes (Bács–Kiskun County)

The excavation of the site was carried out by Zsolt Gallina and Sándor Varga between 1996 and 2002 [14]. The cemetery, which consisted of 206 graves, is divided into northern and southern parts, based on the type and the orientation of the graves, the grave goods, and the position of the arms. The northern part is further split into east and west sides, based on the density of the graves. Due to the characteristics of the soil, the shape of the graves has been preserved well: several grave pits with sidewall niches have been found, as well as traces of other Avar-age burial customs, which were quite rare in the 10th century (e.g., patterns of post holes in the grave pit). The archeological findings include jewelry and clothing ornaments, such as hoops around the head (e.g., S-terminalled and penannular hair rings, earrings), neck jewelry (e.g., neck rings, beads), arm jewelry (e.g., bracelets, rings), and dress fittings, as well as weapons (e.g., archery equipment, axes) and implements (e.g., fire-lighting equipment, knives).

Based on the composition of the findings, the cemetery dates back to the period from the first third of the 10th century to the first third of the 11th century. It is believed that the first generation may still have belonged to the conquerors and the last generation may have been buried there during the reign of King (St.) Stephen I. According to a former classification based on the size of the cemetery and the composition of the findings, the cemetery was described as a "commoner cemetery". In a more recent classification, the cemetery belongs to the 10th century village cemetery group.

The state of preservation of the anthropological material is generally good or medium, excluding the sub-adult skeletons. During the anthropological examinations, 136 adult and 50 sub-adult remains were distinguished, of which 63 were described as male and 83 as female. Based on the distribution of taxonomic features, in addition to the predominance of Europids (88%), a smaller proportion of components belonging to the Mongoloid type were also present (9.3%). Based on biological distance calculations, the population of the cemetery shows similarities to other 10th century and Avar-era series.

Appendix A.2.2. Ibrány–Esbóhalom (Szabolcs–Szatmár–Bereg County)

The excavation of the cemetery was carried out between 1985 and 1990 under the leadership of Eszter Istvánovits [13]. The remains of 274 individuals were found in 269 graves. Based on the finds, the 10th century part of the cemetery can be well characterized by the low number of burials with weapon- and horse-riding-related grave goods and dress fittings (e.g., pendant mounts), as well as simple wire jewelry (e.g., bracelets) and implements (e.g., knives, fire-lighting equipment). Within the 10th century part, a group with different ethnicities was also distinguished, based on the grave orientation, burial customs, and findings. In addition to the coins related to the reign of the kings of the Árpád dynasty, the burials of the 11th century were characterized by S-terminalled hair rings, beads, and rings. In virtue of the archeological material, continuity was assumed between the two parts of the site. The use of the cemetery dates back to 940–1075. Based on the number of the graves and the composition of the archeological material, it was characterized as a commoner cemetery. The state of preservation of the anthropological material is medium, often poor. In anthropological studies, 98 men, 82 women, 74 children (inf I–II), and 20 young (juvenile) individuals have been distinguished; thus there is a male surplus of adults. Taxonomic analysis revealed a predominance of Europeans; Mongoloid traits were observed in four individuals. Craniometric analysis showed a discrepancy between the 10th and 11th century parts, raising the question that the two cemetery parts may hide different populations.

Appendix A.2.3. Magyarhomorog–Kónyadomb (Hajdú–Bihar County)

The systematic excavation of the cemetery was started by István Dienes between 1961 and 1971 and was finished by László Kovács between 1985 and 1988 [15]. The cemetery has a "pagan" segment with 17 graves from the 10th century characterized by a high number of burials with weapon- and horse-riding-related grave goods and a significant male surplus. The larger part consists of 523 "Christian" graves of the 11–12th centuries in which the gender rate is more balanced. The cemetery is one of the few sites where the burial custom of giving tomb furnishings, which was forbidden by the Christian liturgy, can be observed after the turn of the millennium and the adoption of Christianity (e.g., weapon- and horse-riding-related grave goods in the tombs dated with coins related to the reign of the Árpád dynasty kings). Different types of jewelry were the most common archeological findings: hoops around the head (e.g., S-terminalled hair rings), neck jewelry (e.g., beads), bracelets, and various types of rings. Based on the location of the tombs containing coins, the cemetery started from a central core and expanded evenly toward its edges. During the archeological analysis, the possibility of discontinuity between the two parts of the cemetery arose. Therefore, archeologists suggest that the cemetery separately has a 10th century so-called quarter cemetery and a 11–12th century village cemetery. During the anthropological analysis, 11 men, 3 women, and 3 children of unknown sex were identified in the 10th century part. The state of preservation was generally of medium quality. Based on the taxonomic analysis, the Europid groups dominated (about 60%), but, overall, the proportion of those showing Europo–Mongoloid traits is significant with one individual being classified as Mongoloid type. In paleopathological alterations, developmental abnormalities predominated, which address several further questions about kin relationships within the group. Concerning the 11–12th century village cemetery, the state of preservation of the skeletons was generally low. A total of 126 males, 174 females, 187 adolescents (infantia I–II), and 36 juvenile individuals were distinguished. According to the craniometric analysis, the dolichocranial skulls were dominant, and the taxonomic analysis revealed that skulls with Europid characteristics were in the highest number, but there were also Europo–Mongoloid characteristics and in six cases Mongolid types were present (the state of preservation highly limited the classification). In our analysis, we examined whether the "Christian" part could have been contiguous with the "pagan" part, i.e., it hides descendants of the same population or its population originates from elsewhere. Special attention should also be paid to the possible connections of the previously described cemeteries of Karos and Kenézlő of a similar age.

Appendix A.2.4. Nagytarcsa–Homokbánya (Pest County)

In 1967, under the leadership of László Kovács 21 graves were excavated and there are data on another 7 disturbed graves, but the cemetery can only be considered as partially excavated (estimated at about 40–50 graves) [49]. The poor archeological findings consisted of penannular hair rings, twisted neck rings, wire bracelets, rings, and ball buttons. Two burials with weapon-related grave goods (archery equipment, an ax) and four burials with horse-riding-related deposits (e.g., pear-shaped stirrups) were excavated in the cemetery. On the grounds of the findings, the cemetery was dated to the second half of the 10th century. The state of preservation of the anthropological material is moderate, often low. The anthropological analysis [50] identified the skeletons of 8 males, 15 females, and 4 unspecified children. Five skulls belonging to the Europid type proved to be suitable for taxonomic analysis.

Appendix A.2.5. Püspökladány–Eperjesvölgy (Hajdú–Bihar County)

The excavation was carried out under the leadership of Ibolya M. Nepper and Márta Sz. Máthé between 1977 and 1982 [11]. A total of 637 graves were found in the cemetery, but due to double burials the remains of 641 individuals were identified. Based on the findings, the cemetery can be divided into two parts. The "pagan" western part (about one-third of the cemetery) dating back to the 10th century is characterized by burials with weapon- (e.g., archery equipment, sabers, swords) and horse-riding-related (pear-shaped stirrups) grave goods, as well as jewelry, such as hoops around the head (e.g., penannular hair rings), necklaces (e.g., beads), and arm/hand jewelry (e.g., bracelets, rings), dress fittings, and implements (e.g., knives, fire-lighting equipment). The other part was likely used after the adoption of Christianity in the 11th century, based on the more common occurrence of coin-dated burials and relatively late grave-good types (e.g., twisted and braided rings, foil beads, S-terminalled hair rings). On the strength of the size of the cemetery and the composition of the archeological material, this cemetery belongs to the commoner cemeteries in the former classification and to the 10–11th century village cemeteries in the more recent classification. The anthropological material is of medium or poor preservation. During the anthropological analysis of the cemetery [11,51], 191 male, 163 female, and 256 sub-adult (unspecified sex) individuals were described. According to studies on craniometric and body height data, continuity was assumed between the 10th century and 11th century parts of the cemetery.

Appendix A.2.6. Sárrétudvari–Hízóföld (Hajdú–Bihar County)

The site was excavated between 1980 and 1985 under the leadership of Ibolya M. Nepper [11]. The site, with 262 graves, is considered the largest 10th century cemetery in Hungary. The cemetery contains a very high proportion of burials with weapon- (archery equipment, sabers, axes) and horse-riding-related (e.g., pear-shaped stirrups) grave goods, and the archeological findings consist of jewelry, such as hoops around the head (penannular hair rings), neck jewelry (neck rings, beads), and arm jewelry (e.g., bracelets, beads), dress fittings, and implements (e.g., knives, fire-lighting equipment). Based on the composition of the findings and the lack of coins and grave goods dated to the reign of the kings of the Árpád dynasty, the cemetery can be dated to the 10th century. Formerly it was classified as a commoner cemetery, and in the new classification it belongs to the 10th century village cemeteries.

The skeletal remains are of good/medium preservation. During the extensive anthropological analysis (e.g., [52]), 265 individuals were determined, of whom 98 belonged to sub-adult and 162 to adult categories. Based on the skulls suitable for taxonomic studies, the series shows European characteristics with the presence of Cromagnoid and Nordoid elements.

Appendix A.2.7. Szegvár–Oromdűlő (Csongrád County)

The site contained 372 graves which were excavated under the leadership of Gábor Lőrinczy between 1980/1983 and 1996 [53], but many burials (about 75–85) were destroyed due to previous disturbances. Five additional graves were excavated 30–40 m away from the tight array of the cemetery. The archeological material is characterized by jewelry, such as hoop jewelry around the head (penannular hair rings, coiled hair rings, S-terminalled hair rings, hoops with spiral pendants), neck rings, bracelets, and rings, and, less frequently, implements (knives, fire-lighting equipment) and dress fittings (one grave) were excavated. The use of the cemetery dates back to the period between the second third of the 10th century and the middle of the 11th century. Based on its size and the composition of the grave goods, the cemetery was classified formerly as a commoner cemetery of the 10–11th centuries and recently as a 10–11th century village cemetery.

In anthropological studies [54], skeletons of 110 males, 114 females, and 148 sub-adults of indeterminate sex were described. During the taxonomic analysis, the predominance of Europid type skulls (Cromagnoid, Nordoid) was detected in both the 10th and 11th century groups, but a small proportion of Mongoloid features were also described. Based on the craniometric data, it is assumed that a population change occurred at the turn of the 10–11th centuries.

Appendix A.2.8. Szegvár–Szőlőkalja (Csongrád County)

The site of 62 graves containing the burials of 63 individuals was excavated in 1979 by Katalin Hegedűs [55]. A burial of opposite orientation (E–W) was found at a distance of 30–35 m from the cemetery array. The poor archeological findings consist of penannular hair rings, beads, wire bracelets, ball buttons, and knives. Horse-riding-related equipment (a fragment of a bridle) and weapons (arrowheads) were unearthed in one burial. Based on the composition of the archeological material, the site dates back to the 10th century, and it was classified formerly as a commoner cemetery and recently as a 10th century villager cemetery.

During the analysis of anthropological findings [55], 25 male and 25 female skeletons were described. Based on taxonomic studies, the population composition is heterogenous; the individuals belonged to the Europid, Mongoloid, and Europo–Mongoloid types.

Appendix A.2.9. Vörs–Papkert (Somogy County)

The cemetery was excavated between 1983 and 1993 under the leadership of László Költő, Szilvia Honti, and József Szentpéteri [16]. The cemetery as a whole is still unpublished, but it is dated back to the turn of the 8–9th centuries to the turn of the 10–11th centuries. The 716 excavated burials are mostly from the Late Avar and Carolingian periods, but sporadic burials of 33 people can be dated to the time of the Hungarian conquest. Therefore, the study of the cemetery could help investigate the supposed survival of the Transdanubian Late Avar population through the 9–10th centuries. Extensive anthropological and serological investigations were carried out at the cemetery [56], but contradictory results were obtained (e.g., concerning the sex determination).

References

1. Róna-Tas, A. *Hungarians and Europe in the Early Middle Ages: An Introduction to Early Hungarian History*; CEU Press: Budapest, Hungary, 1999; ISBN 978-963-9116-48-1.
2. Szabados, G. The origins and the transformation of the early Hungarian state. In *Reform and Renewal in Medieval East and Central Europe: Politics, Law and Society*; Miljan, S., Simon, A., Halász, B.É., Eds.; Knjižnici Hrvatske Akademije Znanosti i Umjetnosti: Zagreb, Croatia, 2019; pp. 9–30. ISBN 9786060380023.
3. Hampel, J. *Újabb tanulmányok a honfoglalási kor emlékeiről*; MTA: Budapest, Hungary, 1907. (In Hungarian)
4. Szőke, B. *A Honfoglaló és kora Árpád-kori magyarság régészeti emlékei*; Gerevich, L., Erdélyi, I., Kutizán, I.B., Párducz, M., Patek, E., Salamon, Á., Eds.; Akadémiai Kiadó: Budapest, Hungary, 1967. (In Hungarian)
5. Csősz, A.; Szécsényi-Nagy, A.; Csakyova, V.; Langó, P.; Bódis, V.; Köhler, K.; Tömöry, G.; Nagy, M.; Mende, B.G. Maternal Genetic Ancestry and Legacy of 10th Century AD Hungarians. *Sci. Rep.* **2016**, *6*. [CrossRef]
6. Neparáczki, E.; Maróti, Z.; Kalmár, T.; Kocsy, K.; Maár, K.; Bihari, P.; Nagy, I.; Fóthi, E.; Pap, I.; Kustár, Á.; et al. Mitogenomic data indicate admixture components of Central-Inner Asian and Srubnaya origin in the conquering Hungarians. *PLoS ONE* **2018**, *13*, e0205920. [CrossRef]
7. Neparáczki, E.; Maróti, Z.; Kalmár, T.; Maár, K.; Nagy, I.; Latinovics, D.; Kustár, Á.; Pálfi, G.; Molnár, E.; Marcsik, A.; et al. Y-chromosome haplogroups from Hun, Avar and conquering Hungarian period nomadic people of the Carpathian Basin. *Sci. Rep.* **2019**, *9*, 16569. [CrossRef] [PubMed]
8. Tömöry, G.; Csányi, B.; Bogácsi-Szabó, E.; Kalmár, T.; Czibula, Á.; Csősz, A.; Priskin, K.; Mende, B.; Langó, P.; Downes, C.S.; et al. Comparison of Maternal Lineage and Biogeographic Analyses of Ancient and Modern Hungarian Populations. *Am. J. Phys. Anthropol.* **2007**, *134*, 354–368. [CrossRef] [PubMed]
9. Neparáczki, E.; Juhász, Z.; Pamjav, H.; Fehér, T.; Csányi, B.; Zink, A.; Maixner, F.; Pálfi, G.; Molnár, E.; Pap, I.; et al. Genetic structure of the early Hungarian conquerors inferred from mtDNA haplotypes and Y-chromosome haplogroups in a small cemetery. *Mol. Genet. Genom.* **2017**, *292*, 201–214. [CrossRef] [PubMed]
10. QGIS Development Team QGIS Geographic Information System. Open Source Geospatial Foundation Project. 2020. Available online: http://qgis.osgeo.org/ (accessed on 13 March 2020).
11. Nepper, I.M. *Hajdú-Bihar megye 10–11. századi sírleletei I-II*; Bende, I.K., Ed.; Déri Múzeum–Magyar Nemzeti Múzeum–Magyar Tudományos Akadémia Régészeti Intézete: Budapest, Hungary; Debrecen, Hungary, 2002; ISBN 963 9046 79. (In Hungarian)

12. Neparáczki, E.; Kocsy, K.; Tóth, G.E.; Maróti, Z.; Kalmár, T.; Bihari, P.; Nagy, I.; Pálfi, G.; Molnár, E.; Raskó, I.; et al. Revising mtDNA haplotypes of the ancient Hungarian conquerors with next generation sequencing. *PLoS ONE* **2017**, *12*, e0174886. [CrossRef]
13. Istvánovics, E. *A Rétköz honfoglalás és Árpád-kori emlékanyaga*; Kovács, L., Révész, L., Eds.; Jósa András Múzeum–Magyar Nemzeti Múzeum–Magyar Tudományos Akadémia Régészeti Intézete: Nyíregyháza, Hungary, 2003; ISBN 9637220488. (In Hungarian)
14. Gallina, J.Z.; Varga, S. *A Duna-Tisza közének Honfoglalás és kora Árpád-kori temetői, sír- és kincsleletei I. A Kalocsai Sárköz a 10–11. században [Magyarország honfoglalás kori és kora Árpád-kori sírleletei 10.]*; Magyar Nemzeti Múzeum/MTA BTK Régészeti Intézet/SZTE BTK Régészet Tanszék/Viski Károly Múzeum: Szeged, Hungary; Budapest, Hungary, 2016; ISBN 9789633064954. (In Hungarian)
15. Kovács, L. *Magyarhomorog-Kónya-domb 10. századi szállási és 11-12. századi falusi temetője*; MARTIN OPITZ: Szeged, Hungary; Budapest, Hungary, 2019; ISBN 9789639987463. (In Hungarian)
16. Költő, L.; Szentpéteri, J. A Vörs-Papkert "B" lelőhely VIII–XI. századi temetője. In *Évezredek üzenete a láp világából. Régészeti kutatások a Kis-Balaton területén 1979–1992*; Költő, L., Vándor, L., Eds.; Somogy Megyei Múzeumok Igazgatósága—Zala Megyei Múzeumok Igazgatósága: Kaposvár, Hungary; Zalaegerszeg, Hungary, 1996; pp. 115–121. (In Hungarian)
17. Meyer, M.; Kircher, M. Illumina sequencing library preparation for highly multiplexed target capture and sequencing. *Cold Spring Harb. Protoc.* **2010**, *5*. [CrossRef] [PubMed]
18. Kircher, M.; Sawyer, S.; Meyer, M. Double indexing overcomes inaccuracies in multiplex sequencing on the Illumina platform. *Nucleic Acids Res.* **2012**, *40*, 1–8. [CrossRef]
19. Rohland, N.; Harney, E.; Mallick, S.; Nordenfelt, S.; Reich, D. Partial uracil—DNA—glycosylase treatment for screening of ancient DNA. *Phil. Trans. R. Soc. B* **2015**, *370*, 20130624. [CrossRef] [PubMed]
20. Maricic, T.; Whitten, M.; Pääbo, S. Multiplexed DNA sequence capture of mitochondrial genomes using PCR products. *PLoS ONE* **2010**, *5*, e14004. [CrossRef]
21. Martin, M. Cutadapt removes adapter sequences from high-throughput sequencing reads. *EMBnet J.* **2011**, *17*, 10–12. [CrossRef]
22. Andrews, S. FastQC A Quality Control tool for High Throughput Sequence Data. 2016. Available online: http://www.bioinformatics.babraham.ac.uk/projects/fastqc/ (accessed on 16 April 2017).
23. Andrews, R.M.; Kubacka, I.; Chinnery, P.F.; Lightowlers, R.N.; Turnbull, D.M.; Howell, N. Reanalysis and revision of the Cambridge reference sequence for human mitochondrial DNA. *Nat. Genet.* **1999**, *23*, 147. [CrossRef]
24. Li, H.; Durbin, R. Fast and accurate short read alignment with Burrows-Wheeler transform. *Bioinformatics* **2009**, *25*, 1754–1760. [CrossRef] [PubMed]
25. Li, H.; Handsaker, B.; Wysoker, A.; Fennell, T.; Ruan, J.; Homer, N.; Marth, G.; Abecasis, G.; Durbin, R. The Sequence Alignment/Map format and SAMtools. *Bioinformatics* **2009**, *25*, 2078–2079. [CrossRef] [PubMed]
26. Broad Institute Picard Tools. 2016. Available online: https://broadinstitute.github.io/picard/ (accessed on 16 April 2017).
27. Jónsson, H.; Ginolhac, A.; Schubert, M.; Johnson, P.L.F.; Orlando, L. mapDamage2.0: Fast approximate Bayesian estimates of ancient DNA damage parameters. *Bioinformatics* **2013**, *29*, 1682–1684. [CrossRef] [PubMed]
28. Renaud, G.; Slon, V.; Duggan, A.T.; Kelso, J. Schmutzi: Estimation of contamination and endogenous mitochondrial consensus calling for ancient DNA. *Genome Biol.* **2015**, *16*, 224. [CrossRef]
29. Weissensteiner, H.; Pacher, D.; Kloss-Brandstätter, A.; Forer, L.; Specht, G.; Bandelt, H.-J.; Kronenberg, F.; Salas, A.; Schönherr, S. HaploGrep 2: Mitochondrial haplogroup classification in the era of high-throughput sequencing. *Nucleic Acids Res.* **2016**, *44*, W58–W63. [CrossRef]
30. Skoglund, P.; Storå, J.; Götherström, A.; Jakobsson, M. Accurate sex identification of ancient human remains using DNA shotgun sequencing. *J. Archaeol. Sci.* **2013**. [CrossRef]
31. Csáky, V.; Gerber, D.; Szeifert, B.; Egyed, B.; Stégmár, B.; Botalov, S.G.; Grudochko, I.V.; Matveeva, N.P.; Zelenkov, A.S.; Sleptsova, A.V.; et al. Early medieval genetic data from Ural region evaluated in the light of archaeological evidence of ancient Hungarians. *Sci. Rep.* **2020**. [CrossRef]
32. R Core Development Team R: A Language and Environment for Statistical Computing, 3.2.1. 2015. Available online: http//www.r-project.org (accessed on 17 April 2017).
33. Excoffier, L.; Lischer, H.E.L. Arlequin suite ver 3.5: A new series of programs to perform population genetics analyses under Linux and Windows. *Mol. Ecol. Resour.* **2010**, *10*, 564–567. [CrossRef]
34. Slatkin, M. A measure of population subdivision based on microsatellite allele frequencies. *Genetics* **1995**, *139*, 457–462. [CrossRef]
35. Derenko, M.; Malyarchuk, B.; Denisova, G.; Perkova, M.; Litvinov, A.; Grzybowski, T.; Dambueva, I.; Skonieczna, K.; Rogalla, U.; Tsybovsky, I.; et al. Western Eurasian ancestry in modern Siberians based on mitogenomic data. *BMC Evol. Biol.* **2014**, *14*, 217. [CrossRef]
36. Kivisild, T. Maternal ancestry and population history from whole mitochondrial genomes. *Investig. Genet.* **2015**, *6*, 3. [CrossRef] [PubMed]
37. Allentoft, M.E.; Sikora, M.; Sjögren, K.-G.; Rasmussen, S.; Rasmussen, M.; Stenderup, J.; Damgaard, P.B.; Schroeder, H.; Ahlström, T.; Vinner, L.; et al. Population genomics of Bronze Age Eurasia. *Nature* **2015**, *522*, 167–172. [CrossRef] [PubMed]
38. De Barros Damgaard, P.; Marchi, N.; Rasmussen, S.; Peyrot, M.; Renaud, G.; Korneliussen, T.; Moreno-Mayar, J.V.; Pedersen, M.W.; Goldberg, A.; Usmanova, E.; et al. 137 ancient human genomes from across the Eurasian steppes. *Nature* **2018**, *557*, 369–374. [CrossRef]

39. Jeong, C.; Wang, K.; Wilkin, S.; Taylor, W.T.T.; Miller, B.K.; Bemmann, J.H.; Stahl, R.; Chiovelli, C.; Knolle, F.; Ulziibayar, S.; et al. A Dynamic 6,000-Year Genetic History of Eurasia's Eastern Steppe. *Cell* **2020**. [CrossRef]
40. Olajos, T. Az avar továbbélés kérdéséről. *Tiszatáj* **2001**, 50–56. (In Hungarian). Available online: https://docplayer.hu/4510605-Az-avar-tovabbeles-kerdeserol.html (accessed on 8 December 2020).
41. Takács, M. A honfoglalás kor és a településrégészet. In *Magyar őstörténet*; Vásáry, I., Fodor, P., Eds.; MTA BTK Történettudományi Intézet: Budapest, Hungary, 2014; pp. 137–150. ISBN 978-963-9627-87-1. (In Hungarian)
42. Révész, L. *A 10–11. századi temetők regionális jellemzői a Keleti-Kárpátoktól a Dunáig [Magyarország honfoglalás kori és kora Árpád-kori sírleletei 13]*; Szegedi Tudományegyetem Régészeti Tanszéke–Magyar Tudományos Akadémia Régészeti Intézete–Magyar Nemzeti Múzeum–Martin Opitz Kiadó: Budapest, Hungary; Debrecen, Hungary, 2020; ISBN 9789639987609. (In Hungarian)
43. Steuer, H. Frühgeschichtliche Sozialstrukturen in Mitteleuropa. Zur Analyze der Auswertungsmethode des archäologischen Quellenmaterials. In *Geschichtswissenschaft und Archäologie. Untersuchungen zur Siedlungs-, Wirtschafts- und Kirchengeschichte*; Jankuhn, H., Wenskus, R., Eds.; Verlag, J.T.: Stuttgart, Germany, 1979; pp. 595–633. ISBN 3799566228. (In German)
44. Brather, S. "Etnikai értelmezés" és struktúratörténeti magyarázat a régészetben. *Korall* **2006**, *7*, 23–72. (In Hungarian)
45. Bóna, I. A honfoglaló magyarság régészeti hagyatékának társadalomtörténeti tanulságai. *Magy. Tudomány* **1997**, *104*, 1451–1461. (In Hungarian)
46. Kovács, L. A Kárpát-medence honfoglalás és kora Árpád-kori szállási és falusi temetői. Kitekintéssel az előzményekre. In *A honfoglalás kor kutatásának legújabb eredményei: Tanulmányok Kovács László 70. születésnapjára*; Révész, L., Wolf, M., Eds.; Szegedi Tudományegyetem Régészeti Tanszéke–Magyar Tudományos Akadémia Magyar Őstörténeti Témacsoport–Martin Opitz Kiadó: Szeged, Hungary, 2013; pp. 511–604. ISBN 978 963 306 241 8. (In Hungarian)
47. Harke, H. *Angelsächsische Waffengräber des 5. bis 7. Jahrhunderts*; Rheinland-Verlag and Habelt: Köln, Germany; Bonn, Germany, 1992; ISBN 3792712172. (In German)
48. Harke, H. The nature of burial data. In *Burial & Society: The Chronological and Social Analysis of Archaeological Burial Data*; Jensen, C.K., Nielsen, K.H., Eds.; University Press: Aarhus, Danmark, 1997; pp. 19–27. ISBN 8772886862/9788772886862.
49. Kovács, L. Honfoglalás kori sírok Nagytarcsán II: A homokbányai temetőrészlet. Adatok a nyéltámaszos balták, valamint a trapéz alakú kengyelek értékeléséhez. *Commun. Archaeol. Hung.* **1986**, *13*, 93–121. (In Hungarian)
50. Lotterhof, E. The Anthropological Investigation of the Tenth Century Population Excavated at Nagytarcsa. *Anthropol. Hung.* **1973**, *12*, 41–61.
51. Lenkey, Z.; Szathmáry, L.; Csóri, Z.; János, I.; Csoma, E.; Medveczky, Z. Tizenöt 8–13. századi népesség kraniológiai elemzése. In *Árpád előtt, Árpád után. Antropológiai vizsgálatok az Alföld I–XIII. századi csontvázleletein*; Szathmáry, L., Ed.; JATE Press: Szeged, Hungary, 2008; pp. 24–70. (In Hungarian)
52. Oláh, S. Sárrétudvari–Hízóföld Honfoglalás kori Temetőjének Történeti Embertani Értékelése. Ph.D. Thesis, József Attila University, Szeged, Hungary, 1990. (In Hungarian).
53. Bende, L.; Lőrinczy, G. A szegvár-oromdűlői 10–11. századi temető. In *A Móra Ferenc Múzeum Évkönyve—Studia Archaeologica 3*; Bende, L., Lőrinczy, G., Szalontai, C., Eds.; Csongrád Megyei Múzeumok Igazgatósága: Szeged, Hungary, 1997; pp. 201–285. (In Hungarian)
54. Marcsik, A. Szegvár-Oromdűlő 10. és 11. századi embertani leleteinek vizsgálata. In *A Móra Ferenc Múzeum Évkönyve—Studia Archaeologica, 3*; Bende, L., Lőrinczy, G., Szalontai, C., Eds.; Csongrád Megyei Múzeumok Igazgatósága: Szeged, Hungary, 1997; pp. 287–322. (In Hungarian)
55. Lőrinczy, G. Szegvár–Szőlőkalja X. századi temetője. Communicationes Archaeologicae Hungariae. *Commun. Archaeol. Hung.* **1985**, 141–162. (In Hungarian) ISBN:HU ISSN 0231-13.
56. Költő, L.; Szentpéteri, J.; Bernert, Z.; Papp, I. Families, finds and generations: An interdisciplinary experiment at the early medieval cemetery of Vörs–Papkert B. In *Mensch, Siedlung und Landschaft im Wechsel der Jahrtausende am Balaton. People, Settlement and Landscape on Lake Balaton over the millennia. Castellum Pannonicum Pelsonense (CPP4)*; Heinrich-Tamáska, O., Straub, P., Eds.; Verlag Marie Leidorf GmbH: Rahden, Germany, 2014; pp. 361–390. ISBN 978-3-89646-154-4.

Article

First Bronze Age Human Mitogenomes from Calabria (Grotta Della Monaca, Southern Italy)

Francesco Fontani [1,†], Elisabetta Cilli [1,†], Fabiola Arena [2,3], Stefania Sarno [4,*], Alessandra Modi [5], Sara De Fanti [4,6], Adam Jon Andrews [1,4], Adriana Latorre [1], Paolo Abondio [4], Felice Larocca [3,7], Martina Lari [5], David Caramelli [5], Emanuela Gualdi-Russo [2] and Donata Luiselli [1,*]

1 Department of Cultural Heritage, University of Bologna, via Degli Ariani 1, 48121 Ravenna, Italy; francesco.fontani2@unibo.it (F.F.); elisabetta.cilli@unibo.it (E.C.); a.andrews@unibo.it (A.J.A.); adriana.latorre@studio.unibo.it (A.L.)
2 Department of Neuroscience and Rehabilitation, University of Ferrara, Corso Ercole I D'Este 32, 44121 Ferrara, Italy; fabiola.arena@unife.it (F.A.); emanuela.gualdi@unife.it (E.G.-R.)
3 Centro Regionale di Speleologia "Enzo dei Medici", via Lucania 3, 87070 Roseto Capo Spulico, Italy; felicelarocca1964@gmail.com
4 Department of Biological Geological and Environmental Sciences, University of Bologna, via Selmi 3, 40126 Bologna, Italy; sara.defanti2@unibo.it (S.D.F.); paolo.abondio2@unibo.it (P.A.)
5 Department of Biology, Università degli Studi di Firenze, via del Proconsolo 12, 50125 Firenze, Italy; alessandra.modi@unifi.it (A.M.); martina.lari@unifi.it (M.L.); david.caramelli@unifi.it (D.C.)
6 Interdepartmental Centre "Alma Mater Research Institute on Global Challenges and Climate Change (Alma Climate)", University of Bologna, via Petroni 26, 40126 Bologna, Italy
7 Gruppo di Ricerca Speleo-Archeologica, Università degli Studi di Bari "Aldo Moro", Piazza Umberto I 1, 70121 Bari, Italy
* Correspondence: stefania.sarno2@unibo.it (S.S.); donata.luiselli@unibo.it (D.L.)
† These authors equally contributed to the article.

Citation: Fontani, F.; Cilli, E.; Arena, F.; Sarno, S.; Modi, A.; De Fanti, S.; Andrews, A.J.; Latorre, A.; Abondio, P.; Larocca, F.; et al. First Bronze Age Human Mitogenomes from Calabria (Grotta Della Monaca, Southern Italy). *Genes* **2021**, *12*, 636. https://doi.org/10.3390/genes12050636

Academic Editor: Michael Hofreiter

Received: 26 February 2021
Accepted: 20 April 2021
Published: 25 April 2021

Abstract: The Italian peninsula was host to a strong history of migration processes that shaped its genomic variability since prehistoric times. During the Metal Age, Sicily and Southern Italy were the protagonists of intense trade networks and settlements along the Mediterranean. Nonetheless, ancient DNA studies in Southern Italy are, at present, still limited to prehistoric and Roman Apulia. Here, we present the first mitogenomes from a Middle Bronze Age cave burial in Calabria to address this knowledge gap. We adopted a hybridization capture approach, which enabled the recovery of one complete and one partial mitochondrial genome. Phylogenetic analysis assigned these two individuals to the H1e and H5 subhaplogroups, respectively. This preliminary phylogenetic analysis supports affinities with coeval Sicilian populations, along with Linearbandkeramik and Bell Beaker cultures maternal lineages from Central Europe and Iberia. Our work represents a starting point which contributes to the comprehension of migrations and population dynamics in Southern Italy, and highlights this knowledge gap yet to be filled by genomic studies.

Keywords: ancient DNA; paleogenomics; human; mitochondrial DNA; archaeology; Italy; Bronze Age

1. Introduction

The genomic variability within the Italian Peninsula is greater today than in any European country, which may suggest that this area played a pivotal role in the peopling of the Mediterranean in the past [1]. Genetic studies based on autosomal and uniparental markers [2–4], genome-wide [1,5–9] and whole genome approaches [10] have dissected the clinal variability of the present Italian population, revealing multilayered patterns of prehistorical and historical processes of migration and admixture that occurred throughout the Peninsula. In particular, these studies provided evidence of an early divergence of Italian groups dating back to the Late Glacial period, with further differentiation attributed to Neolithic and Bronze Age migrations [10].

Modern Southern Italian populations have been extensively studied [4,6,7,10,11], highlighting a long series of migration processes and cultural exchanges that influenced this area. High frequencies of maternal lineages from the Caucasus and the Levant were retrieved, which predate the Neolithic and may support the role of this area as a refugium during the Last Glacial Maximum (LGM) [6,12]. A wide genetic Mediterranean "*continuum*" was identified, which links Southern Italy with Crete and the Caucasus following Neolithic and Bronze Age migrations from the Near East. In addition, during the Bronze Age, a non-steppe contribution derived from the Caucasus was detected [7,10].

The Metal Age, in particular, deeply characterized the complex population processes of protohistoric Italy, leaving clear signatures in modern populations [13]. Territorialization spread in both Tyrrhenian and Adriatic populations during the Bronze Age, and dynamism was further promoted following novel social structuring [14]. The emergence of elite groups during the second millennium BC [15] resulted in community organization pivoted around kinship and inherited rank in Southern communities [16]. The Cosenza province (central Calabria) played an important role in the landscape of prehistoric studies of Southern Italy due to this structuring. Evidence of consistent mining activities nearby the Sila Plateau allowed protohistoric Calabria to establish itself as one of the fundamental territories for metallurgy in Italy, and to connect this area in a strong exchange relationship within the lower Tyrrhenian coastal communities [17]. The presence of a Protoapennine culture in many Middle Bronze Age (MBA) Calabrian contexts, such as Broglio di Trebisacce [18], was suggested by Ardesia [19] to be the "Calabrian" Rodì-Tindari-Vallelunga (RTV) horizon, thus underlining the strong interactions and shared cultural patterns between the southern areas and the western islands of the peninsula. Nevertheless, a deep analysis of funerary contexts from protohistoric Southern Italy is essential. The presence of numerous cave sites that have not yet been investigated in the area of Tyrrhenian Calabria prompt the need for a deeper knowledge of the ancient communities of Southern Italy. The northwestern sector of Calabria is rich in cave sites, with over two hundred caves present along the Tyrrhenian coast and inland, such as Grotte di Cirella in Diamante, Grotte di Torre Talao in Scalea, Grotta della Madonna in Praia a Mare, Grotta del Romito in Papasidero, Grotte di Sant'Angelo in Cassano allo Ionio and Grotta di Donna Marsilia [20]. Therefore, paleogenetic studies from this area, in particular, are warranted to investigate early population dynamics. To complement genetic studies on modern populations, ancient DNA (aDNA) studies have the potential to provide precise insights into early heritage and migrations, coupled with sociocultural aspects, of past societies. However, aDNA studies in Southern Italy are limited to prehistoric [21,22] and Roman [23] Apulia. Hence, we aimed here to produce mitogenome data on Bronze Age communities from Calabria, in Southern Italy, to provide new information on population dynamics from this understudied area and to contextualize them with archaeological and anthropological evidence. To achieve this, we explored the remains of a MBA cave burial in Tyrrhenian Calabria, Grotta della Monaca (Figure 1).

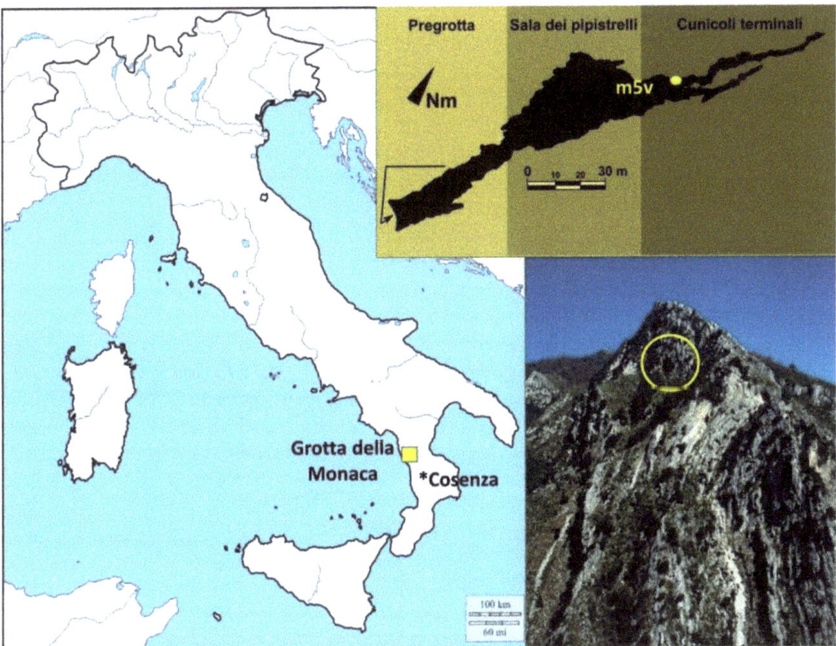

Figure 1. Location of Grotta della Monaca. The entrance of the cave is marked by a yellow circle on the bottom right image. The scheme of the cave with the *m5v* area is represented at the top right.

2. Materials and Methods

2.1. Archaeological Context and Human Remains

Grotta della Monaca is a karstic cave located in Sant'Agata di Esaro at 600 m above sea level, close to the Esaro river (Cosenza, northwestern Calabria, Southern Italy) (Figure 1). The cave has a long human history [20,24,25] due to mining activities that began during the Upper Paleolithic period and intensified during the Late Neolithic/Early Chalcolithic period [26]. The cave partially changed its role during the MBA, when the deepest areas were used as burial grounds [27,28]. Archaeological investigations conducted between 2003 and 2010 recovered a mass grave with numerous human skeletal remains in a niche of the area of Cunicoli terminali (hereafter *m5v*). Collective burials in a natural cave were a common and uninterrupted ritual in protohistoric Southern Italy, particularly in the coastal sites facing the Tyrrhenian Sea [29]. From the Early Bronze Age (EBA), this funerary practice was reserved for high-status individuals [30]. Fragmented pottery and funerary artifacts evidence a link between the burials in Grotta della Monaca and the so-called Apennine culture [31], which was an archaeological complex that was widespread in the Italian Peninsula and Sicily during the second half of the second millennium BC. The remains found in the *m5v* context represent at least 24 different commingled individuals (File S1 and Table S1), and were in a poor state of preservation due to several taphonomic factors [32]. The osteological remains buried in *m5v* of Grotta della Monaca were examined by traditional anthropological methodologies at the Laboratory of Archaeo-Anthropology and Forensic Anthropology of the University of Ferrara (See File S1 for details about the methods used and the anthropological data acquired).

Radiocarbon analysis of the human bone samples was conducted at CEDAD (Center of Applied Physics, Dating and Diagnostics, University of Salento, Lecce, Italy). Six out of seven samples were successfully dated and were assigned to the MBA period. Calibrated and uncalibrated dates are summarized in Table S2.

2.2. Experimental Procedure

Ancient DNA analysis was performed on seven individuals (Table S3 and Figure S1) from the *m5v* mass grave. Due to the poor preservation of the remains, the only petrous bone available (preferred because it is DNA-rich [33,34]) was associated with individual 24. Thus, four shafts of long bones (GdM5, GdM6, GdM7, GdM12) and two cranial fragments (GdM11, GdM14) were sampled along with the petrous bone (GdM24). The samples were selected based on the preservation state, and took into account the bone elements used to calculate the minimum number of individuals (MNI), to make sure we sampled different individuals. Analysis was conducted in the ancient DNA Laboratory (aDNA Lab) of the Department of Cultural Heritage, University of Bologna (Ravenna, Italy). Post-PCR procedures were performed in a separate laboratory, and negative control was carried out alongside each experiment to confirm the absence of intra-laboratory contamination. All materials and instruments used were DNA-free and decontaminated before use with bleach and/or DNA ExitusPlus™ (AppliChem GmbH, Darmstadt, Germany) and researchers followed high-sterility protocols according to ancient DNA authenticity criteria [35–38].

Specimens were first cleaned with sterile scalpels to remove sediments and calcareous concretions, and 1–2 mm of the outer layer of bones was abraded with circular drilling discs. Bones were decontaminated for 1 h under UV irradiation, and 140 to 340 mg fragments of the bone specimens were pulverized, while 255 mg of bone powder was collected from petrous bone by directly drilling the densest region of the cochlea.

2.3. Ancient DNA Extraction

DNA from the samples was extracted following a two-day silica spin-column protocol [39,40] with slight modifications, as in Cilli et al. [41]. Samples were digested in a rotator at 37 °C overnight in 3000 µL solution composed of 2700 µL EDTA, 37.5 µL Proteinase K and H_2O. Supernatant was centrifuged and transferred with PB binding buffer (Qiagen, Hilden, Germany) into silica columns (Roche-High Pure Viral Nucleic Acid Large Volume Kit, Roche, Basel, Switzerland). After two washing steps with PE buffer (Qiagen), DNA was eluted in 50 µL of EB buffer (Qiagen).

2.4. Library Preparation and Enrichment

Double-stranded library preparation was performed according to Carøe et al. [42]. Libraries were indexed with Illumina sequencing adapters and no enzymatic damage repair was carried out in order to preserve patterns of ancient DNA fragments. A qPCR quantification was performed prior to indexing, in order to assign correct amplification cycles to each library and to reduce the formation of heteroduplex structures. A total volume of 30 µL from each library was split in 2/3 aliquots and used for indexing PCR. PCR indexing products were purified with MinElute PCR Purification kit (Qiagen), eluted in EBT buffer (10 mM Tris-Cl, pH 8.5 and 0.05% Tween-20) and analyzed on Agilent 2100 Bioanalyzer (Agilent, Santa Clara, CA, USA) to produce qualitative and quantitative estimations of the libraries.

Indexed libraries were equimolarly pooled to a total of maximum 2 µg and enriched for human mitochondrial DNA (mtDNA) for the bead-capture method using long-range PCR products, according to Maricic et al. [43]. The captured libraries were quantified using Agilent 2100 Bioanalyzer and sequenced on Illumina MiSeq platform at the Advanced Genomic Centre of the University of Florence. The sequencing was run as paired-end with $75 \times 2 + 8 + 8$ cycles.

2.5. Bioinformatic Analysis

Raw reads were first visualized with FastQC [44], and adapter sequences were filtered using AdapterRemoval v2.3.0 [45]. Fragments shorter than 30 bp and with a '*minquality* < *25*' (PHRED values) were discarded, and paired-end reads with overlaps of at least 10 bp were collapsed into single sequences. Filtered reads were mapped against the revised Cambridge Reference Sequence (rCRS, NC_012920.1) [46] with BWA v.0.7.17-r188 [47].

Command *'aln'* was set with -n number to 0.01, seed-length option deactivated and -o value at 3 for tolerating a higher number of gaps in the alignment. The generated .sai file was aligned against the reference fasta file with *'samse'* command, and then converted to .bam files and sorted by leftmost coordinate using SAMtools v1.10 [48] *'view'* and *'sort'* commands. The Picard tool MarkDuplicates [49] was used on the .bam files to mark and remove PCR duplicates, and reads were locally realigned through Genome Analysis Toolkit tools' [50] *'RealignerTargetCreator'* and *'IndelRealigner'* commands, to minimize the number of mismatches around the indels that may be easily mistaken as SNPs.

Data authenticity and post-mortem damage were investigated through MapDamage v2.0.8 [51] and a new rescaled .bam file was generated with the *'rescale'* option, that downscales the quality scores at positions likely affected by damage patterns. Damage patterns at the 5′ and 3′ ends of the reads were also computed with contDeam, a tool provided within the Schmutzi package [52], and present-day human contamination estimates were performed with Schmutzi, using a non-redundant database of 256 mitogenomes available in the software package. Results are summarized in Table S3. Mitochondrial genomes were called using Schmutzi, and variants were then investigated using GATK *'HaplotypeCaller'* and filtered with *'VariantFiltration'* command, storing only detected variants with quality *'QUAL20'* (≥20) and depth of coverage *'COV3'* (≥3). Diagnostic variants were checked and verified with snpToolkit v2.2.3 [53], and only variants with ratio −r 0.9 were used to determine the haplogroup.

2.6. Phylogenetic Analyses

Phylogenetic analyses were performed on the complete mitogenome of individual GdM24, while GdM7 was only analyzed for haplogroup assignment. The other samples were not included due to inconsistencies in the data (Table S3). The mtDNA haplogroup assignment was predicted using both Haplofind [54] and HaploGrep [55] software, and checked manually according to PhyloTree build 17 [56]. The consensus sequence obtained for GdM24 was then compared to a reference dataset of 97 ancient individuals extracted from the literature (Table S4). In addition, 13 published mitogenomes from ancient African and European samples that supply basal lineage information for phylogenetic reconstruction were included in the analyses (Table S5). The RSRS and rCRS reference sequences were also added for comparison purposes in the Network analysis.

Sequence alignment was performed with the DNA Alignment software [57] and was visually confirmed. The poly-C and AC-indels at 303–315, 515–524, 573–576 and 16,180–16,193 positions, as well as the nucleotide position 16,519, were excluded from phylogenetic analyses [58]. A Median Joining Network was calculated by using Network v.10.2 [59] with no pre- or post-processing steps. Furthermore, a phylogenetic tree, based on the same multi-alignment dataset after retaining unique sequences, was reconstructed with BEAST v1.8.0 [60] by using reference ancient dated samples as calibration points (Table S5). The best substitution model for the dataset was tested with Mega 5.2 [61], resulting in the Hasegawa–Kishino–Yano model with fixed fraction of invariable sites and gamma distributed rates (HKY + I + G). We ran BEAST analysis with 200,000,000 MCMC generations, and sampled every 2000 iterations by testing different combinations of clock models (i.e., Strict clock vs. Uncorrelated Relaxed Lognormal-ULN clock) and tree models (i.e., Constant Population Size vs. Bayesian Skyline). Chain convergence was assessed with Tracer v 1.6, resulting in Effective Sampling Size (ESS) values higher than 200 for all the parameters and in all the tested model combinations. Evaluating the maximum likelihood estimates (MLE) for the four combinations of clock and tree models further revealed the best support for the Relaxed ULN clock model and Bayesian Skyline tree model; therefore, this combination was chosen for the final step of analysis. The Maximum Clade Credibility tree was calculated using TreeAnnotator v1.8.0 by discarding the first 20% of iterations as burn-in. The resulting phylogenetic tree was graphically represented with FigTree [62].

3. Results

Six human bone fragments and one petrous bone from the *m5v* burial area of Grotta della Monaca were analyzed for ancient mtDNA on the Illumina MiSeq platform. Five out of seven samples contained low concentrations of DNA (Table S3) and were not further analyzed. For the sample GdM24, we obtained 10,471 mapping reads with quality ≥30, which resulted in a mean coverage of 34X, and 100% of positions covered at least 3-fold. GdM7 contained 550 reads mapped with quality ≥30, resulting in a mean coverage of 2X, and 91% of positions covered at least 1-fold. Analysis of the read datasets with MapDamage v2.0.8 revealed miscoding lesion distribution patterns typical of ancient DNA in both samples (Figure S2). The presence of positions' specific substitutions increases at the ends of the reads, with an average frequency of 28.67% (GdM7) and 45.46% (GdM24) of C-to-T substitution at the first base at 5′, and similar average frequency of 33.80% (GdM7) and 42.85% (GdM24) G-to-A substitution at 3′ ends. The average fragment length is around 56.49 for GdM24 and 77.33 for GdM7. These data are consistent with the antiquity of the samples and the environmental conditions of the site.

The GdM24 mitochondrial genome was assigned with 85.63% accuracy (Haplogrep) and 1.0 score (Haplofind) to the H1e mitochondrial subclade, based on the relevance of the eleven diagnostic variants detected (263G, 1438G, 3010A, 4769G, 5460A, 6779G, 8860G, 15326G, 16172C, 16224C, 16519C). Regarding GdM7, haplogroup assignment was inferred on the partial sequence, evaluating the presence of diagnostic variants (263G, 1438G, 2626C, 4769G, 8860G, 16304C, 16310A). This appears to belong to the mitochondrial subclade H5, with 79.95% accuracy on Haplogrep and 1.0 quality score on Haplofind, respectively. However, due to the low coverage and incompleteness of the mitogenome, GdM7 was not included in the subsequent phylogenetic analysis.

Phylogenetic reconstruction performed with both Median Joining Network (MJN) and Bayesian Evolutionary Analysis of Sampling Trees (BEAST) confirmed the attribution of GdM24 to the mtDNA haplogroup H1e. Accordingly, the sample indeed clusters with the other H1e samples extracted from the literature, within a clade that splits up after H1 and the other basal-considered lineages (Figures 2 and 3). A star-like pattern characterized both H1 and H1e haplotypes in the Network analysis (Figure 2). In particular, the GdM24 sample branches, along with two Sicilian Bronze Age samples (I3876 and I7774 [63]) from a median vector, originated from the H1e basal node which contains samples from Middle/Late Neolithic and Bronze Age cultures from Central Europe, and a Bell Beaker sample from Southern Italy (I4930 [63]). The same clustering pattern was consistently confirmed in the BEAST reconstructed phylogenetic tree (Figure 3), with the branch leading to the GdM24 sequence, particularly, dating at 4125 years BP (95% HPD: 2301–6332).

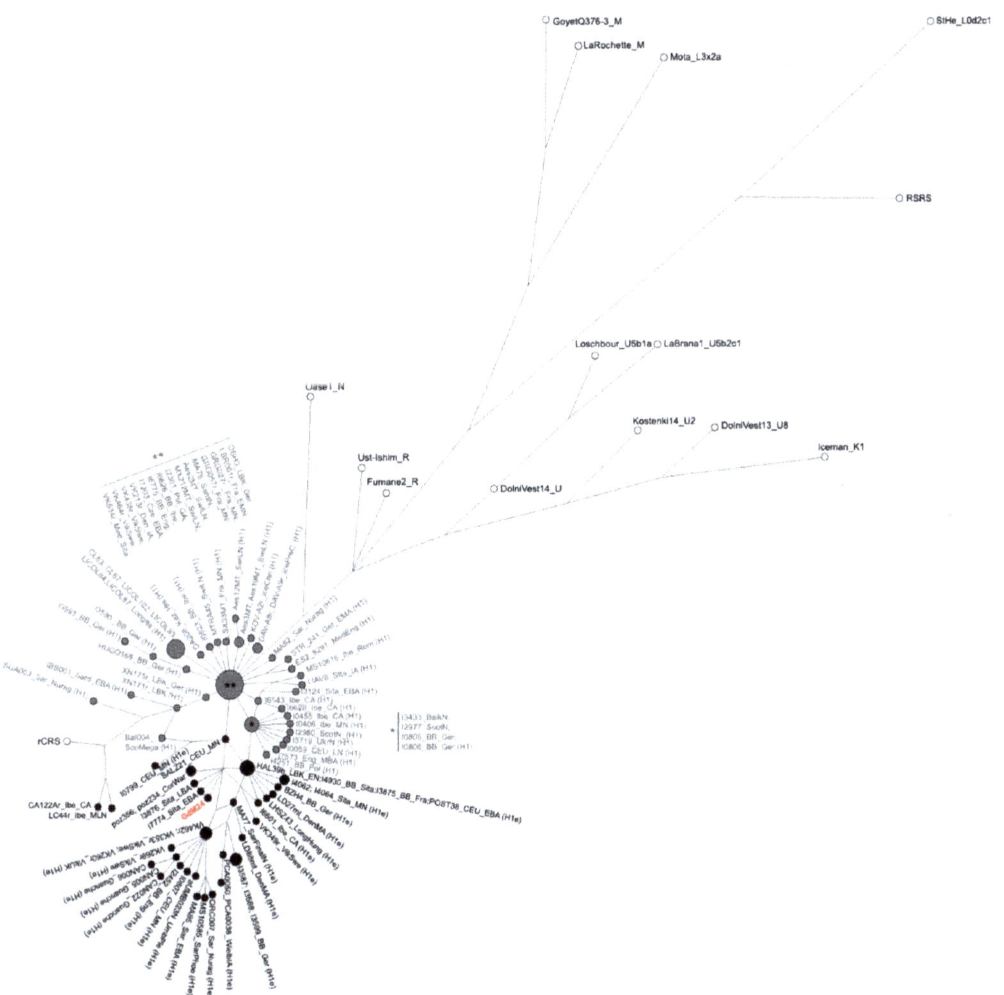

Figure 2. Median Joining Network based on the complete GdM24 mitogenome (shown in red) compared to a dataset of ancient reference (in white), H1 (grey) and H1e (black) mitogenomes (see Table S4 for dataset).

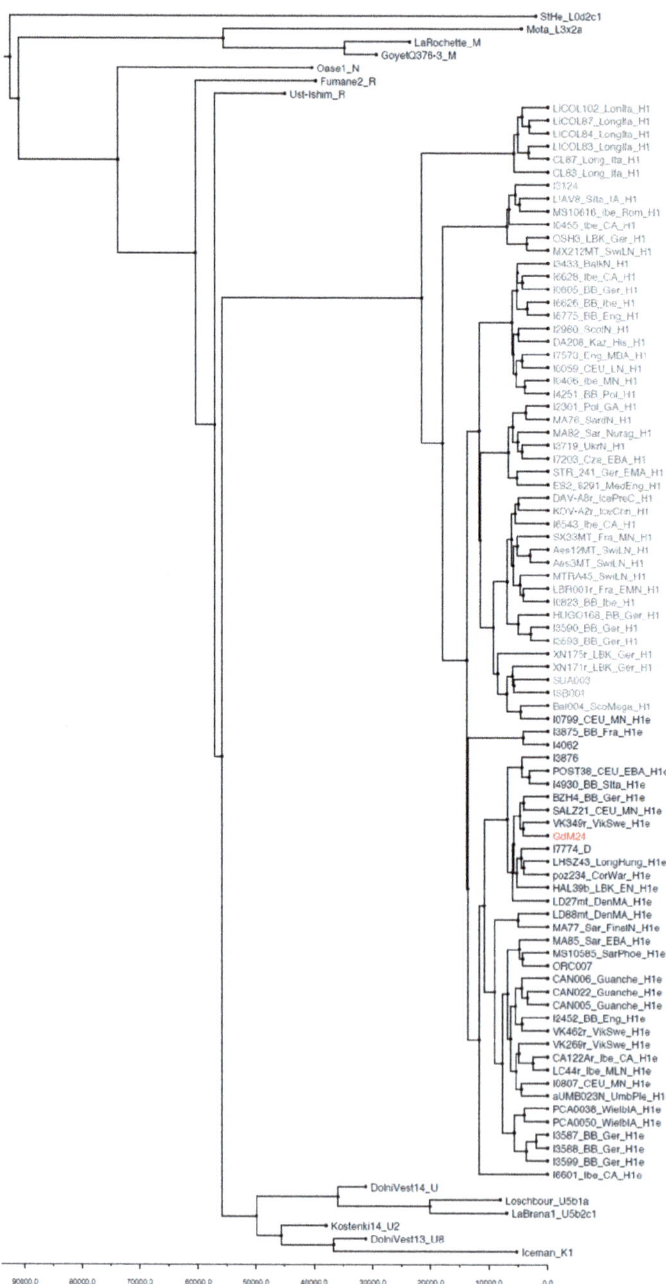

Figure 3. Phylogenetic tree reconstructed with BEAST using dated ancient reference samples as calibration points (Tables S4 and S5).

4. Discussion

Excluding Sicily and Sardinia that experienced different processes of migration and admixture and were covered by several paleogenomic studies [63–69], Bronze Age Southern Italy has not been studied with an aDNA approach. In this study we attempted analysis of the mitochondrial DNA variation of seven MBA samples from Grotta della Monaca cave, located in Calabria, Southern Italy, with the aim of filling the gap in the contribution of ancient genetic data to the population dynamics of this peculiar area. This is of particular interest and importance considering the key role that the Italian Peninsula played in past demographic processes, as evidenced by the wide variability in genomic diversity in present-day Italian populations [1].

Despite the poor preservation of the remains, the hybridization capture method allowed us to retrieve and reconstruct the complete mitochondrial genome of individual GdM24, and partially reconstruct the mitochondrial genome of individual GdM7. These are the first attested ancient molecular data from the Calabria region, as highlighted by a recent review of the state-of-the-art European paleogenomic data [70]. Unsurprisingly, GdM24 was the only sample for which the petrous bone was available from this burial area. Previous studies suggest that this skeletal element preserves the endogenous DNA better than other bones or tissues [33,71]. Thus, this type of sample is confirmed as fundamental in the recovery of ancient endogenous DNA from particularly difficult contexts, such as those at low latitudes.

The two samples for which we obtained complete mitogenomes are both assigned to mitochondrial macrohaplogroup H. Nowadays, this is the most widespread in Europe, encompassing over 40% of the total mtDNA variation, and could be considered as the quintessential Eurasiatic marker. Previous studies have shown how haplogroup H arrived in Europe from the Near East before LGM (~22,000 BP), survived in south western glacial refugia and then colonized Central and Northern Europe [72,73]. In particular, GdM24 was assigned to mitochondrial sublineage H1e, and GdM7 to H5.

Among the subclades, H1 is the most frequent in modern-day Europeans, followed by H3 and H5 [74]. When compared to other present-day European and Middle Eastern populations, haplogroup H1 shows frequency peaks in the Franco-Cantabrian region, declining from the west towards Eastern and Southern Europe [75]. Taking into account both the distribution and the coalescence age of this subhaplogroup (9.3–12.8 kya; [76]), it has been hypothesized that it could be correlated with a Late Glacial re-expansion of populations from the Franco-Iberian refugial areas at the end of the Ice Age, from ~15 kya, or at the end of the Younger Dryas, ~11.5 kya [77–81]. Along with H3, H1 subclade was also proposed as a marker for a Bell Beaker complex expansion originating in Iberia [72]; however, a recent genome-wide study on ancient genomes does not support migration as an important mechanism of the spread of this culture between Iberia and Central Europe [66]. Ancient DNA evidence also highlighted the earliest presence of the H1e subhaplogroup in Neolithic samples from Hungary (~7000 BP) [82] and Germany (~6100 BP) [72]. In Italy, the H1e subhaplogroup was retrieved in Sicily from two Middle Neolithic samples (~6800 BP), an Eneolithic (~4700 BP) and an EBA sample (~4000 BP) [63]. In the present study, phylogenetic analyses of GdM24 highlighted the proximity of this sample, among others, with two Sicilian Bronze Age samples (I3876 and I7774 [63]) recovered in Marcita and Contrada Paolina necropolises, respectively.

Regarding H5, this subhaplogroup shows a coalescence time of ~10.7–17.1 kya [76]. There is no agreement about the place of origin of H5; hypotheses have been made about eastern Mediterranean [74,80] or south-west European origin, where it has expanded after the Ice Age, with several possible dispersal routes [76]. However, the dynamics that led to the diffusion of this haplogroup are not yet clearly understood. The oldest evidence of H5 lineage was retrieved in Anatolia (~8200 BP) [83] and Bulgaria (~7600 BP) [84]. A migration with a center of origin in the eastern Mediterranean, that would have carried this lineage into Italy and, to a lesser extent, the western Mediterranean, until northern and western Iberia, has been hypothesized [80].

The mitochondrial haplogroups found in present-day Calabria ([7], personal data) include all of the haplogroups that have been found in the ancient individuals analyzed here (H1e and H5). None of these are unique to the ancient Southern Italian populations and, therefore, cannot be taken as indisputable proof of local population continuity. Nonetheless, they deserve to be more deeply analyzed at the nuclear level to obtain more information about the ancestry and migration patterns of this area.

In the neighboring Sicily, past population dynamics are better understood due to sampling conducted in two projects [63,66] that detected ancestry typical of early European farmers as a mixture of Anatolia Neolithic and Western Hunter-Gatherer, and no evidence of steppe ancestry in Middle Neolithic samples. They found evidence of steppe ancestry in the EBA, by around 2200 BC, with the forming of a clade with EBA Mallorca, suggesting that the population may harbor ancestry most plausibly from Iberia, with a scenario of west-to-east gene flow. Fernandes et al. [63] retrieved Iranian-related ancestry in Sicily, from the MBA (1800–1500 BC) and Late Bronze Age (LBA), which was widespread among the Aegean by the MBA, along with the Mycenaean cultural expansion or earlier.

The active role of Tyrrhenian Calabria in cultural connections between eastern Sicily, the Aeolian Islands and the Adriatic environments in protohistoric times have been highlighted [85–89]. These connections are demonstrated by the chrono-typological study of pottery, and reflect strong relations between central-southern Calabria, the Aeolian Islands and the northern part of Sicily [90]. However, the so-called Palma Campania *facies* that identifies the MBA contexts of southwestern Italy still generates uncertainties, especially in Tyrrhenian Calabria, due to the lack of in-depth knowledge of the archaeological contexts, and the absence of ancient genomic data. In light of this evidence, our data represents a baseline for a deeper comprehension of the population pattern of prehistoric and protohistoric Southern Italy.

In the area of Tyrrhenian Calabria, the presence of over two hundred cave sites with attested past human frequentation, not yet investigated, prompts the need for a deeper knowledge of the ancient communities of Southern Italy. Future studies, focused on genome-wide data from this peculiar area, could clarify migratory and demographic processes that took place in the prehistory and protohistory of the Southern European continent, provide information about the ancestry of individuals through time and allow the study of intrinsic differences in migratory behavior related to sex-biased processes, adaptation and admixture.

Supplementary Materials: The following are available online at https://www.mdpi.com/article/10.3390/genes12050636/s1, File S1: Archaeological context and anthropological analyses, Figure S1: specimens used for molecular analyses, Figure S2: MapDamage plot, Table S1: Summary of biological profile and pathologies in individuals from Grotta della Monaca involved in genetic analyses, Table S2: Radiocarbon dating and calibration executed on human samples from Grotta della Monaca, Table S3: Summary of the genomic analyses, Table S4: Ancient samples belonging to H1 and H1e haplogroups used in phylogenetic analyses; Table S5: Dataset of reference ancient sequences used as tip calibration points in the BEAST analyses.

Author Contributions: Conceptualization, D.L., E.G.-R. and F.L.; methodology E.C., F.F., and A.M.; software F.F., S.S., and A.J.A.; validation E.C., and F.F.; formal analysis E.C., F.F., F.A., A.L., A.J.A., and S.D.F.; investigation E.C., F.F., and F.A.; resources D.L.; data curation F.F., and S.S.; writing—original drafting preparation, F.F., E.C. and F.A.; writing—review and editing F.F., E.C., F.A., S.D.F., S.S., A.M., A.J.A., A.L., P.A., D.C., M.L., F.L., E.G.-R., and D.L.; visualization S.S. and F.F.; supervision D.L.; project administration D.L.; funding acquisition D.L. All authors have read and agreed to the published version of the manuscript.

Funding: This work was supported by the MIUR-PRIN 20177PJ9XF grant to D.L. and D.C. A.J.A. is supported by SeaChanges—European Union's Horizon 2020 programme, under the Marie Skłodowska-Curie grant agreement No 813383.

Institutional Review Board Statement: Not applicable.

Informed Consent Statement: Not applicable.

Data Availability Statement: Analyzed data for GdM24 sample are available on NCBI GenBank (Accession number: MW853672). Raw data for all samples have been uploaded on ENA database (Study accession number: PRJEB44125).

Acknowledgments: We would like to thank Thomas Mignani, Andrea De Giovanni, Irene Boreggio and Isabella Cavaliere who assisted during the experimental procedures.

Conflicts of Interest: The authors declare no conflict of interest.

References

1. Sazzini, M.; Ruscone, G.A.G.; Giuliani, C.; Sarno, S.; Quagliariello, A.; De Fanti, S.; Boattini, A.; Gentilini, D.; Fiorito, G.; Catanoso, M.; et al. Complex interplay between neutral and adaptive evolution shaped differential genomic background and disease susceptibility along the Italian peninsula. *Sci. Rep.* **2016**, *6*, 32513. [CrossRef]
2. Capelli, C.; Brisighelli, F.; Scarnicci, F.; Arredi, B.; Caglia', A.; Vetrugno, G.; Tofanelli, S.; Onofri, V.; Tagliabracci, A.; Paoli, G.; et al. Y chromosome genetic variation in the Italian peninsula is clinal and supports an admixture model for the Mesolithic–Neolithic encounter. *Mol. Phylogenet. Evol.* **2007**, *44*, 228–239. [CrossRef]
3. Brisighelli, F.; Álvarez-Iglesias, V.; Fondevila, M.; Blanco-Verea, A.; Carracedo, Á.; Pascali, V.L.; Capelli, C.; Salas, A. Uniparental Markers of Contemporary Italian Population Reveals Details on Its Pre-Roman Heritage. *PLoS ONE* **2012**, *7*, e50794. [CrossRef]
4. Boattini, A.; Martinez-Cruz, B.; Sarno, S.; Harmant, C.; Useli, A.; Sanz, P.; Yang-Yao, D.; Manry, J.; Ciani, G.; Luiselli, D.; et al. Uniparental Markers in Italy Reveal a Sex-Biased Genetic Structure and Different Historical Strata. *PLoS ONE* **2013**, *8*, e65441. [CrossRef]
5. Di Gaetano, C.; Voglino, F.; Guarrera, S.; Fiorito, G.; Rosa, F.; Di Blasio, A.M.; Manzini, P.; Dianzani, I.; Betti, M.; Cusi, D.; et al. An Overview of the Genetic Structure within the Italian Population from Genome-Wide Data. *PLoS ONE* **2012**, *7*, e43759. [CrossRef]
6. Sarno, S.; Boattini, A.; Carta, M.; Ferri, G.; Alù, M.; Yao, D.Y.; Ciani, G.; Pettener, D.; Luiselli, D. An Ancient Mediterranean Melting Pot: Investigating the Uniparental Genetic Structure and Population History of Sicily and Southern Italy. *PLoS ONE* **2014**, *9*, e96074. [CrossRef]
7. Sarno, S.; Boattini, A.; Pagani, L.; Sazzini, M.; De Fanti, S.; Quagliariello, A.; Ruscone, G.A.G.; Guichard, E.; Ciani, G.; Bortolini, E.; et al. Ancient and recent admixture layers in Sicily and Southern Italy trace multiple migration routes along the Mediterranean. *Sci. Rep.* **2017**, *7*, 1984. [CrossRef]
8. Fiorito, G.; Di Gaetano, C.; Guarrera, S.; Rosa, F.; Feldman, M.W.; Piazza, A.; Matullo, G. The Italian genome reflects the history of Europe and the Mediterranean basin. *Eur. J. Hum. Genet.* **2016**, *24*, 1056–1062. [CrossRef]
9. Raveane, A.; Aneli, S.; Montinaro, F.; Athanasiadis, G.; Barlera, S.; Birolo, G.; Boncoraglio, G.; Di Blasio, A.M.; Di Gaetano, C.; Pagani, L.; et al. Population structure of modern-day Italians reveals patterns of ancient and archaic ancestries in Southern Europe. *Sci. Adv.* **2019**, *5*, eaaw3492. [CrossRef]
10. Sazzini, M.; Abondio, P.; Sarno, S.; Gnecchi-Ruscone, G.A.; Ragno, M.; Giuliani, C.; De Fanti, S.; Ojeda-Granados, C.; Boattini, A.; Marquis, J.; et al. Genomic history of the Italian population recapitulates key evolutionary dynamics of both Continental and Southern Europeans. *BMC Biol.* **2020**, *18*, 51. [CrossRef]
11. Sarno, S.; Petrilli, R.; Abondio, P.; De Giovanni, A.; Boattini, A.; Sazzini, M.; De Fanti, S.; Cilli, E.; Ciani, G.; Gentilini, D.; et al. Genetic history of Calabrian Greeks reveals ancient events and long term isolation in the Aspromonte area of Southern Italy. *Sci. Rep.* **2021**, *11*, 3045. [CrossRef]
12. De Fanti, S.; Barbieri, C.; Sarno, S.; Sevini, F.; Vianello, D.; Tamm, E.; Metspalu, E.; Van Oven, M.; Hübner, A.; Sazzini, M.; et al. Fine Dissection of Human Mitochondrial DNA Haplogroup HV Lineages Reveals Paleolithic Signatures from European Glacial Refugia. *PLoS ONE* **2015**, *10*, e0144391. [CrossRef] [PubMed]
13. Capocasa, M.; Anagnostou, P.; Bachis, V.; Battaggia, C.; Bertoncini, S.; Biondi, G.; Boattini, A.; Boschi, I.; Brisighelli, F.; Caló, C.M.; et al. Linguistic, geographic and genetic isolation: A collaborative study of Italian populations. *J. Anthropol. Sci* **2014**, *92*, 201–231. [PubMed]
14. Cavazzuti, C.; Skeates, R.; Millard, A.R.; Nowell, G.; Peterkin, J.; Bernabò Brea, M.; Cardarelli, A.; Salzani, L. Flows of people in villages and large centres in Bronze Age Italy through strontium and oxygen isotopes. *PLoS ONE* **2019**, *14*, e0209693. [CrossRef]
15. Cavazzuti, C.; Arena, A. The Bioarchaeology of Social Stratification in Bronze Age Italy. *Arheo* **2020**, *37*, 69–106.
16. Iacono, F. *Archaeology of Late Bronze Age Interaction and Mobility at the Gates of Europe: People, Things … and Networks around the Southern Adriatic Sea*; Bloomsbury Academic: London, UK; New York, NY, USA, 2019; ISBN 978-1-350-03614-7.
17. Vanzetti, A.; Tinè, V. La Calabria dal Neolitico all'età del ferro. In *Museo dei Brettii e Degli Enotri. Catalogo Dell'esposizione*; Rubbettino: Soveria Mannelli, Italy, 2014; pp. 40–43. ISBN 978-88-498-4233-3. (In Italian)
18. Guzzo, P.G.; Peroni, R.; Bergonzi, G.; Cardarelli, A.; Vagnetti, L. *Ricerche Sulla Protostoria Della SIBARITIDE, 1*; Publications du Centre Jean Bérard: Napoli, Italy, 1982; ISBN 978-2-903189-17-4. (In Italian)
19. Ardesia, V. La cultura di Rodì-Tindari-Vallelunga in Sicilia: Origini, diffusione e cronologia alla luce dei recenti studi. Parte 1. *IpoTESI Preist.* **2014**, *6*, 35–98. (In Italian) [CrossRef]
20. Larocca, F. Grotta Della Monaca (Calabria, Italia Meridionale). Una Miniera Neolitica per l'estrazione Dell'ocra. *Rubricatum Rev. Mus. Gavà* **2012**, *5*, 249–256. (In Italian)
21. Fu, Q.; Posth, C.; Hajdinjak, M.; Petr, M.; Mallick, S.; Fernandes, D.; Furtwängler, A.; Haak, W.; Meyer, M.; Mittnik, A.; et al. The genetic history of Ice Age Europe. *Nature* **2016**, *534*, 200–205. [CrossRef]
22. Posth, C.; Renaud, G.; Mittnik, A.; Drucker, D.G.; Rougier, H.; Cupillard, C.; Valentin, F.; Thevenet, C.; Furtwängler, A.; Wißing, C.; et al. Pleistocene Mitochondrial Genomes Suggest a Single Major Dispersal of Non-Africans and a Late Glacial Population Turnover in Europe. *Curr. Biol.* **2016**, *26*, 827–833. [CrossRef]

23. Emery, M.V.; Duggan, A.T.; Murchie, T.J.; Stark, R.J.; Klunk, J.; Hider, J.; Eaton, K.; Karpinski, E.; Schwarcz, H.P.; Poinar, H.N.; et al. Ancient Roman mitochondrial genomes and isotopes reveal relationships and geographic origins at the local and pan-Mediterranean scales. *J. Archaeol. Sci. Rep.* **2018**, *20*, 200–209. [CrossRef]
24. Quarta, G.; LaRocca, F.; D'Elia, M.; Gaballo, V.; Macchia, M.; Palestra, G.; Calcagnile, L. Radiocarbon Dating the Exploitation Phases of the Grotta Della Monaca Cave in Calabria, Southern Italy: A Prehistoric Mine for the Extraction of Iron and Copper. *Radiocarbon* **2013**, *55*, 1246–1251. [CrossRef]
25. Levato, C.; Larocca, F. The Prehistoric Iron Mine of Grotta Della Monaca (Calabria, Italy). *Anthropol. Praehist.* **2015**, *126*, 25–37.
26. Caricola, I.; Breglia, F.; LaRocca, F.; Hamon, C.; Lemorini, C.; Giligny, F. Prehistoric exploitation of minerals resources. Experimentation and use-wear analysis of grooved stone tools from Grotta della Monaca (Calabria, Italy). *Archaeol. Anthropol. Sci.* **2020**, *12*, 259. [CrossRef]
27. Larocca, F. *La Miniera Pre-Protostorica di Grotta Della Monaca (Sant'Agata di Esaro—Cosenza)*; Centro Regionale di Speleologia "Enzo dei Medici": Roseto Capo Spulico, Italy, 2005. (In Italian)
28. Arena, F.; Larocca, F.; Onisto, N.; Gualdi-Russo, E. *Il Sepolcreto Protostorico Di Grotta Della Monaca in Calabria. Aspetti Antropologici*; Annali Dell'Università di Ferrara: Ferrara, Italy, 2013. (In Italian)
29. Bietti Sestieri, A.M. *L'Italia Nell'età del Bronzo e del Ferro. Dalle Palafitte a ROMOLO (2200-700 a.C.)*; Carocci Editore: Rome, Italy, 2010; ISBN 88-430-5207-1. (In Italian)
30. Cocchi Genick, D. *Criteri di Nomenclatura e di Terminologia Inerente Alla Definizione Delle Forme Vascolari del Neolitico/Eneolitico e del Bronzo/Ferro: Atti del Congresso di Lido di Camaiore, 26–29 Marzo 1998. Vol. 1*; Octavo: Firenze, Italy, 1999; ISBN 978-88-8030-190-5. (In Italian)
31. Larocca, F. Dal Pollino all'Orsomarso. L'uso funerario delle cavità naturali in età pre-protostorica. In *Il Pollino. Barriera Naturale e Crocevia di Culture*; Giornate Internazionali di Archeologia (San Lorenzo Bellizzi, 16–17 Aprile 2016); Colelli, C.E., Larocca, A., Eds.; Collana del Dipartimento di Studi Umanistici, Sezione Archeologia, Università della Calabria: Calabria, Italy, 2018; Volume Ricerche XII, pp. 13–27. (In Italian)
32. Arena, F.; Gualdi-Russo, E. *Taphonomy and Post-Depositional Movements of a Bronze Age Mass Grave in the Archaeological Site of Grotta Della Monaca*; Annali dell'Università di Ferrara: Ferrara, Italy, 2014. (In Italian)
33. Gamba, C.; Jones, E.R.; Teasdale, M.D.; McLaughlin, R.L.; Gonzalez-Fortes, G.; Mattiangeli, V.; Domboróczki, L.; Kővári, I.; Pap, I.; Anders, A.; et al. Genome flux and stasis in a five millennium transect of European prehistory. *Nat. Commun.* **2014**, *5*, 5257. [CrossRef] [PubMed]
34. Hansen, H.B.; Damgaard, P.B.; Margaryan, A.; Stenderup, J.; Lynnerup, N.; Willerslev, E.; Allentoft, M.E. Comparing Ancient DNA Preservation in Petrous Bone and Tooth Cementum. *PLoS ONE* **2017**, *12*, e0170940. [CrossRef] [PubMed]
35. Cooper, A.; Poinar, H.N. Ancient DNA: Do It Right or Not at All. *Science* **2000**, *289*, 1139. [CrossRef]
36. Gilbert, M.T.P.; Bandelt, H.-J.; Hofreiter, M.; Barnes, I. Assessing ancient DNA studies. *Trends Ecol. Evol.* **2005**, *20*, 541–544. [CrossRef]
37. Fulton, T.L. Setting up an Ancient DNA Laboratory. *Methods Mol. Biol.* **2012**, *840*, 1–11. [CrossRef]
38. Fulton, T.L.; Shapiro, B. Setting up an Ancient DNA Laboratory. *Methods Mol. Biol.* **2019**, *1963*, 1–13. [CrossRef]
39. Dabney, J.; Knapp, M.; Glocke, I.; Gansauge, M.-T.; Weihmann, A.; Nickel, B.; Valdiosera, C.; García, N.; Pääbo, S.; Arsuaga, J.-L.; et al. Complete mitochondrial genome sequence of a Middle Pleistocene cave bear reconstructed from ultrashort DNA fragments. *Proc. Natl. Acad. Sci. USA* **2013**, *110*, 15758–15763. [CrossRef]
40. Damgaard, P.B.; Margaryan, A.; Schroeder, H.; Orlando, L.; Willerslev, E.; Allentoft, M.E. Improving access to endogenous DNA in ancient bones and teeth. *Sci. Rep.* **2015**, *5*, 11184. [CrossRef]
41. Cilli, E.; Gabanini, G.; Ciucani, M.M.; De Fanti, S.; Serventi, P.; Bazaj, A.; Sarno, S.; Ferri, G.; Fregnani, A.; Cornaglia, G.; et al. A multifaceted approach towards investigating childbirth deaths in double burials: Anthropology, paleopathology and ancient DNA. *J. Archaeol. Sci.* **2020**, *122*, 105219. [CrossRef]
42. Carøe, C.; Gopalakrishnan, S.; Vinner, L.; Mak, S.S.T.; Sinding, M.H.S.; Samaniego, J.A.; Wales, N.; Sicheritz-Pontén, T.; Gilbert, M.T.P. Single-tube library preparation for degraded DNA. *Methods Ecol. Evol.* **2018**, *9*, 410–419. [CrossRef]
43. Maricic, T.; Whitten, M.; Pääbo, S. Multiplexed DNA Sequence Capture of Mitochondrial Genomes Using PCR Products. *PLoS ONE* **2010**, *5*, e14004. [CrossRef] [PubMed]
44. Andrews, S. FastQC. A Quality Control Tool for High Throughput Sequence Data. Available online: https://www.bioinformatics.babraham.ac.uk/projects/fastqc/ (accessed on 22 April 2021).
45. Schubert, M.; Lindgreen, S.; Orlando, L. AdapterRemoval v2: Rapid adapter trimming, identification, and read merging. *BMC Res. Notes* **2016**, *9*, 88. [CrossRef]
46. Andrews, R.M.; Kubacka, I.; Chinnery, P.F.; Lightowlers, R.N.; Turnbull, D.M.; Howell, N. Reanalysis and revision of the Cambridge reference sequence for human mitochondrial DNA. *Nat. Genet.* **1999**, *23*, 147. [CrossRef]
47. Li, H.; Durbin, R. Fast and accurate short read alignment with Burrows-Wheeler transform. *Bioinformatics* **2009**, *25*, 1754–1760. [CrossRef] [PubMed]
48. Li, H.; Handsaker, B.; Wysoker, A.; Fennell, T.; Ruan, J.; Homer, N.; Marth, G.; Abecasis, G.; Durbin, R.; 1000 Genome Project Data Processing Subgroup. The Sequence Alignment/Map Format and SAMtools. *Bioinformatics* **2009**, *25*, 2078–2079. [CrossRef] [PubMed]
49. Picard Toolkit. Available online: https://broadinstitute.github.io/picard/ (accessed on 22 April 2021).

50. McKenna, A.; Hanna, M.; Banks, E.; Sivachenko, A.; Cibulskis, K.; Kernytsky, A.; Garimella, K.; Altshuler, D.; Gabriel, S.; Daly, M.; et al. The Genome Analysis Toolkit: A MapReduce framework for analyzing next-generation DNA sequencing data. *Genome Res.* **2010**, *20*, 1297–1303. [CrossRef] [PubMed]
51. Jónsson, H.; Ginolhac, A.; Schubert, M.; Johnson, P.L.F.; Orlando, L. mapDamage2.0: Fast approximate Bayesian estimates of ancient DNA damage parameters. *Bioinformatics* **2013**, *29*, 1682–1684. [CrossRef] [PubMed]
52. Renaud, G.; Slon, V.; Duggan, A.T.; Kelso, J. Schmutzi: Estimation of contamination and endogenous mitochondrial consensus calling for ancient DNA. *Genome Biol.* **2015**, *16*, 224. [CrossRef] [PubMed]
53. Namouchi, A. SnpToolkit. Available online: https://github.com/Amine-Namouchi/snpToolkit (accessed on 22 April 2021).
54. Vianello, D.; Sevini, F.; Castellani, G.; Lomartire, L.; Capri, M.; Franceschi, C. HAPLOFIND: A New Method for High-Throughput mtDNA Haplogroup Assignment. *Hum. Mutat.* **2013**, *34*, 1189–1194. [CrossRef] [PubMed]
55. Weissensteiner, H.; Pacher, D.; Kloss-Brandstätter, A.; Forer, L.; Specht, G.; Bandelt, H.-J.; Kronenberg, F.; Salas, A.; Schönherr, S. HaploGrep 2: Mitochondrial haplogroup classification in the era of high-throughput sequencing. *Nucleic Acids Res.* **2016**, *44*, W58–W63. [CrossRef] [PubMed]
56. Van Oven, M. PhyloTree Build 17: Growing the human mitochondrial DNA tree. *Forensic Sci. Int. Genet. Suppl. Ser.* **2015**, *5*, e392–e394. [CrossRef]
57. DNA Alignment Software. Available online: https://www.fluxus-engineering.com/align.htm (accessed on 22 April 2021).
58. Van Oven, M.; Kayser, M. Updated comprehensive phylogenetic tree of global human mitochondrial DNA variation. *Hum. Mutat.* **2009**, *30*, E386–E394. [CrossRef]
59. Network. Available online: https://www.fluxus-engineering.com/sharenet.htm (accessed on 22 April 2021).
60. Drummond, A.J.; Suchard, M.A.; Xie, D.; Rambaut, A. Bayesian Phylogenetics with BEAUti and the BEAST 1.7. *Mol. Biol. Evol.* **2012**, *29*, 1969–1973. [CrossRef]
61. Tamura, K.; Peterson, D.; Peterson, N.; Stecher, G.; Nei, M.; Kumar, S. MEGA5: Molecular Evolutionary Genetics Analysis Using Maximum Likelihood, Evolutionary Distance, and Maximum Parsimony Methods. *Mol. Biol. Evol.* **2011**, *28*, 2731–2739. [CrossRef]
62. Rambaut, A. Figtree v1.4.2, A Graphical Viewer of Phylogenetic Trees. Available online: tree.bio.ed.ac.uk/software/figtree/ (accessed on 22 April 2021).
63. Fernandes, D.M.; Mittnik, A.; Olalde, I.; Lazaridis, I.; Cheronet, O.; Rohland, N.; Mallick, S.; Bernardos, R.; Broomandkhoshbacht, N.; Carlsson, J.; et al. The spread of steppe and Iranian-related ancestry in the islands of the western Mediterranean. *Nat. Ecol. Evol.* **2020**, *4*, 334–345. [CrossRef]
64. Modi, A.; Tassi, F.; Susca, R.R.; Vai, S.; Rizzi, E.; Bellis, G.D.; Lugliè, C.; Gonzalez Fortes, G.; Lari, M.; Barbujani, G.; et al. Complete mitochondrial sequences from Mesolithic Sardinia. *Sci. Rep.* **2017**, *7*, 42869. [CrossRef]
65. Olivieri, A.; Sidore, C.; Achilli, A.; Angius, A.; Posth, C.; Furtwängler, A.; Brandini, S.; Capodiferro, M.R.; Gandini, F.; Zoledziewska, M.; et al. Mitogenome Diversity in Sardinians: A Genetic Window onto an Island's Past. *Mol. Biol. Evol.* **2017**, *34*, 1230–1239. [CrossRef] [PubMed]
66. Olalde, I.; Brace, S.; Allentoft, M.E.; Armit, I.; Kristiansen, K.; Booth, T.; Rohland, N.; Mallick, S.; Szécsényi-Nagy, A.; Mittnik, A.; et al. The Beaker phenomenon and the genomic transformation of northwest Europe. *Nature* **2018**, *555*, 190–196. [CrossRef] [PubMed]
67. Catalano, G.; Lo Vetro, D.; Fabbri, P.F.; Mallick, S.; Reich, D.; Rohland, N.; Sineo, L.; Mathieson, I.; Martini, F. Late Upper Palaeolithic hunter-gatherers in the Central Mediterranean: New archaeological and genetic data from the Late Epigravettian burial Oriente C (Favignana, Sicily). *Quat. Int.* **2020**, *537*, 24–32. [CrossRef]
68. Marcus, J.H.; Posth, C.; Ringbauer, H.; Lai, L.; Skeates, R.; Sidore, C.; Beckett, J.; Furtwängler, A.; Olivieri, A.; Chiang, C.W.K.; et al. Genetic history from the Middle Neolithic to present on the Mediterranean island of Sardinia. *Nat. Commun.* **2020**, *11*, 939. [CrossRef] [PubMed]
69. Modi, A.; Catalano, G.; D'Amore, G.; Di Marco, S.; Lari, M.; Sineo, L.; Caramelli, D. Paleogenetic and morphometric analysis of a Mesolithic individual from Grotta d'Oriente: An oldest genetic legacy for the first modern humans in Sicily. *Quat. Sci. Rev.* **2020**, *248*, 106603. [CrossRef]
70. Olalde, I.; Posth, C. Latest trends in archaeogenetic research of west Eurasians. *Curr. Opin. Genet. Dev.* **2020**, *62*, 36–43. [CrossRef]
71. Pinhasi, R.; Fernandes, D.; Sirak, K.; Novak, M.; Connell, S.; Alpaslan-Roodenberg, S.; Gerritsen, F.; Moiseyev, V.; Gromov, A.; Raczky, P.; et al. Optimal Ancient DNA Yields from the Inner Ear Part of the Human Petrous Bone. *PLoS ONE* **2015**, *10*, e0129102. [CrossRef] [PubMed]
72. Brotherton, P.; Haak, W.; Templeton, J.; Brandt, G.; Soubrier, J.; Jane Adler, C.; Richards, S.M.; Sarkissian, C.D.; Ganslmeier, R.; Friederich, S.; et al. Neolithic mitochondrial haplogroup H genomes and the genetic origins of Europeans. *Nat. Commun.* **2013**, *4*, 1764. [CrossRef]
73. Brandt, G.; Haak, W.; Adler, C.J.; Roth, C.; Szécsényi-Nagy, A.; Karimnia, S.; Möller-Rieker, S.; Meller, H.; Ganslmeier, R.; Friederich, S.; et al. Ancient DNA Reveals Key Stages in the Formation of Central European Mitochondrial Genetic Diversity. *Science* **2013**, *342*, 257–261. [CrossRef]
74. Hernández, C.L.; Dugoujon, J.M.; Novelletto, A.; Rodríguez, J.N.; Cuesta, P.; Calderón, R. The distribution of mitochondrial DNA haplogroup H in southern Iberia indicates ancient human genetic exchanges along the western edge of the Mediterranean. *BMC Genet.* **2017**, *18*, 46. [CrossRef] [PubMed]

75. Álvarez-Iglesias, V.; Mosquera-Miguel, A.; Cerezo, M.; Quintáns, B.; Zarrabeitia, M.T.; Cuscó, I.; Lareu, M.V.; García, O.; Pérez-Jurado, L.; Carracedo, Á.; et al. New Population and Phylogenetic Features of the Internal Variation within Mitochondrial DNA Macro-Haplogroup R0. *PLoS ONE* **2009**, *4*, e5112. [CrossRef] [PubMed]
76. Soares, P.; Achilli, A.; Semino, O.; Davies, W.; Macaulay, V.; Bandelt, H.-J.; Torroni, A.; Richards, M.B. The Archaeogenetics of Europe. *Curr. Biol.* **2010**, *20*, R174–R183. [CrossRef]
77. Torroni, A.; Bandelt, H.-J.; D'Urbano, L.; Lahermo, P.; Moral, P.; Sellitto, D.; Rengo, C.; Forster, P.; Savontaus, M.-L.; Bonné-Tamir, B.; et al. mtDNA Analysis Reveals a Major Late Paleolithic Population Expansion from Southwestern to Northeastern Europe. *Am. J. Hum. Genet.* **1998**, *62*, 1137–1152. [CrossRef]
78. Torroni, A.; Bandelt, H.-J.; Macaulay, V.; Richards, M.; Cruciani, F.; Rengo, C.; Martinez-Cabrera, V.; Villems, R.; Kivisild, T.; Metspalu, E.; et al. A Signal, from Human mtDNA, of Postglacial Recolonization in Europe. *Am. J. Hum. Genet.* **2001**, *69*, 844–852. [CrossRef] [PubMed]
79. Achilli, A.; Rengo, C.; Magri, C.; Battaglia, V.; Olivieri, A.; Scozzari, R.; Cruciani, F.; Zeviani, M.; Briem, E.; Carelli, V.; et al. The Molecular Dissection of mtDNA Haplogroup H Confirms That the Franco-Cantabrian Glacial Refuge Was a Major Source for the European Gene Pool. *Am. J. Hum. Genet.* **2004**, *75*, 910–918. [CrossRef]
80. Pereira, L. High-resolution mtDNA evidence for the late-glacial resettlement of Europe from an Iberian refugium. *Genome Res.* **2005**, *15*, 19–24. [CrossRef] [PubMed]
81. Soares, P.; Ermini, L.; Thomson, N.; Mormina, M.; Rito, T.; Röhl, A.; Salas, A.; Oppenheimer, S.; Macaulay, V.; Richards, M.B. Correcting for Purifying Selection: An Improved Human Mitochondrial Molecular Clock. *Am. J. Hum. Genet.* **2009**, *84*, 740–759. [CrossRef]
82. Lipson, M.; Szécsényi-Nagy, A.; Mallick, S.; Pósa, A.; Stégmár, B.; Keerl, V.; Rohland, N.; Stewardson, K.; Ferry, M.; Michel, M.; et al. Parallel palaeogenomic transects reveal complex genetic history of early European farmers. *Nature* **2017**, *551*, 368–372. [CrossRef]
83. Mathieson, I.; Lazaridis, I.; Rohland, N.; Mallick, S.; Patterson, N.; Roodenberg, S.A.; Harney, E.; Stewardson, K.; Fernandes, D.; Novak, M.; et al. Genome-wide patterns of selection in 230 ancient Eurasians. *Nature* **2015**, *528*, 499–503. [CrossRef]
84. Mathieson, I.; Alpaslan-Roodenberg, S.; Posth, C.; Szécsényi-Nagy, A.; Rohland, N.; Mallick, S.; Olalde, I.; Broomandkhoshbacht, N.; Candilio, F.; Cheronet, O.; et al. The genomic history of southeastern Europe. *Nature* **2018**, *555*, 197–203. [CrossRef] [PubMed]
85. Bernabò Brea, L. *La Sicilia Prima dei Greci*; Il Saggiatore: Milano, Italy, 1961. (In Italian)
86. Bietti Sestieri, A.M. Implicazioni Del Concetto Di Territorio in Situazioni Culturali Complesse: Le Isole Eolie Nell'età Del Bronzo. *Dialoghi Archeol.* **1982**, 2, 39–60. (In Italian)
87. Cazzella, A.; Recchia, G. Malta, Sicily, Aeolian Islands and southern Italy during the Bronze Age: The meaning of a changing relationship. In *Exchange Networks and Local Transformation: Interaction and Local Change in Europe and the Mediterranean from the Bronze Age to the Iron Age*; Oxbow: Oxford, UK, 2013; pp. 80–91.
88. Pacciarelli, M.; Scarano, T.; Crispino, A. The Transition between the Copper and Bronze Ages in Southern Italy and Sicily. In Proceedings of the 7th Archaeological Conference of Central Germany, Halle, Germany, 23–26 October 2014; Landesmuseums für Vorgeschichte Halle: Halle, Germany, 2015.
89. Gori, M.; Revello Lami, M.; Pintucci, A. Editorial: Practices, Representations and Meanings of Human Mobility in Archaeology. *Ex Novo* **2018**, *3*, 1–6. [CrossRef]
90. Marino, S. Across the Strait. New evidence on cultural interconnections and exchanges between Calabria and Sicily during the early Bronze Age. *J. Archaeol. Sci. Rep.* **2020**, *29*, 102164. [CrossRef]

Article

How a Paleogenomic Approach Can Provide Details on Bioarchaeological Reconstruction: A Case Study from the Globular Amphorae Culture

Stefania Vai [1,*], Maria Angela Diroma [1], Costanza Cannariato [1], Alicja Budnik [2], Martina Lari [1], David Caramelli [1] and Elena Pilli [1]

[1] Department of Biology, University of Florence, 50122 Florence, Italy; mariaangela.diroma@unifi.it (M.A.D.); costanza.cannariato@unifi.it (C.C.); martina.lari@unifi.it (M.L.); david.caramelli@unifi.it (D.C.); elena.pilli@unifi.it (E.P.)

[2] Department of Human Biology, Institute of Biological Sciences, Cardinal Stefan Wyszyński University, 01-938 Warsaw, Poland; alicja.budnik.uksw@gmail.com

* Correspondence: stefania.vai@unifi.it

Abstract: Ancient human remains have the potential to explain a great deal about the prehistory of humankind. Due to recent technological and bioinformatics advances, their study, at the palaeogenomic level, can provide important information about population dynamics, culture changes, and the lifestyles of our ancestors. In this study, mitochondrial and nuclear genome data obtained from human bone remains associated with the Neolithic Globular Amphorae culture, which were recovered in the Megalithic barrow of Kierzkowo (Poland), were reanalysed to gain insight into the social organisation and use of the archaeological site and to provide information at the individual level. We were able to successfully estimate the minimum number of individuals, sex, kin relationships, and phenotypic traits of the buried individuals, despite the low level of preservation of the bone samples and the intricate taphonomic conditions. In addition, the evaluation of damage patterns allowed us to highlight the presence of "intruders"—that is, of more recent skeletal remains that did not belong to the original burial. Due to its characteristics, the study of the Kierzkowo barrow represented a challenge for the reconstruction of the biological profile of the human community who exploited it and an excellent example of the contribution that ancient genomic analysis can provide to archaeological reconstruction.

Keywords: ancient DNA; palaeogenomics; Neolithic; multiple burial; kinship; phenotypic traits

Citation: Vai, S.; Diroma, M.A.; Cannariato, C.; Budnik, A.; Lari, M.; Caramelli, D.; Pilli, E. How a Paleogenomic Approach Can Provide Details on Bioarchaeological Reconstruction: A Case Study from the Globular Amphorae Culture. *Genes* **2021**, *12*, 910. https://doi.org/10.3390/genes12060910

Academic Editor: Jennifer A. Leonard

Received: 28 April 2021
Accepted: 8 June 2021
Published: 11 June 2021

1. Introduction

Over the years, the study of ancient DNA study has proven valuable and essential for tracing migrations of historic and prehistoric individuals and groups. Advances in the sequencing and analysis of the genomes of both modern and ancient peoples have facilitated a number of breakthroughs in the understanding of human evolutionary history [1]. In this context, the archaeological excavations and retrieval of skeletal human remains represent a unique opportunity to study a population history and shed light on cultural changes and the lifestyles of our ancestors.

Archaeological excavations in the Żnin district (northwestern Poland), led by Professor Tadeusz Wiślański from the Institute of History and Material Culture of the Polish Academy of Sciences in Poznań, uncovered a Megalithic barrow tomb associated with the Globular Amphora culture (GAC); it contained the remains of several individuals. Through the integration of archaeological, anthropological, and molecular (mtDNA and nuDNA) data, and theories about the development of the GAC, our previous results [2,3] add nuance to the model of Late Neolithic gene flow from the Pontic steppes into Central Europe, showing that the eastern affinities of the GAC in the archaeological record reflect cultural influences rather than movements of people. During the study, archaeogenomic research permitted us to obtain information on past population dynamics, whereas biological traits (such as phenotypic traits) of each individual and relationships within the group in the burial site have not been investigated due to the fact that samples with presumed kinship relations were removed from our previous studies. However, beyond population genetics purposes, as recently proposed, for instance, by [4–7], obtaining more information on the individuals (such as sex estimation and phenotype descriptions of physical appearance and functional traits), and investigating the relationships among them could be important—in general and in particular for the Megalithic barrow of Kierzkowo—for better understanding the social structure and funerary rituals of ancient cultures and preserving and enhancing a specific archaeological site. In fact, the molecular characterisation of human remains can represent an important information source to describe past populations, also for educational purposes: genomic data can be integrated in archaeological and anthropological reconstruction in order to reach a wider public in contexts devoted to science and culture communication, such as museum exhibitions, archaeological parks, and documentaries.

Methods based solely on archaeological and anthropological assessments can sometimes be limited in establishing biological kinship and providing data on biological traits, above all, in contexts such as this Megalithic barrow tomb, in which, over time, anthropic action and post-depositional natural events had a strong impact on the ease of data analysis and interpretation.

Therefore, we decided, here, to reanalyse the molecular data of GAC from our previous studies, attempting to recover information, such as sex, phenotypic, and functional traits to provide individual characterisation, concentrating our attention, in particular, on closely related individuals (previously eliminated from the analysis) and their relations, with archaeological context, to obtain more information useful to reconstruct GAC funerary rituals and social organisation. The complexity of this archaeological site, characterised by a secondary multiple burial of fragmented and mixed human remains belonging to several individuals, represented a challenge for the reconstruction of the biological profile of this community and an excellent example of the contribution that aDNA analysis can provide to archaeological reconstruction. In this context, a molecular investigation turns out to be fundamental to integrate and confirm anthropological and archaeological data/hypotheses, and suggests further insight.

2. Materials and Methods

2.1. The Archaeological Site of Kierzkowo

The archaeological site of Kierzkowo is located in the Żnin district (Kuyavian-Pomeranian voivodeship, northwestern Poland, Figure 1) on top of a hill on the shore of Kierzkowski Lake. Pottery fragments belonging to characteristic decorated globular vessels [8–11], and a radiocarbon date on animal bone found in the site confirmed its attribution to the GAC period (4270 ± 40 BP calibrated to 2896 ± 19 cal BCE) [12].

The original barrow embankment was about 35 m long, 16–18 m wide, and 2 to 4 m high, overgrown by trees and bushes (Figure 1A). After removing the earth embankment, a stone structure approximately 22 m long, W–E, and 3 to 6 m wide, N–S, was revealed (Figure 1B). In the western part of this Megalithic tomb, there was a chamber about 10 m long, about 1.5 m wide, and of the same depth. The chamber was built of stone slabs with a height of about 1 m and a thickness of 0.5 to 0.8 m. Inside, a large boulder divided the chamber into two unequal parts. On the south side, a corridor lined with stones led to the chamber [11,13,14] (Figure 1C).

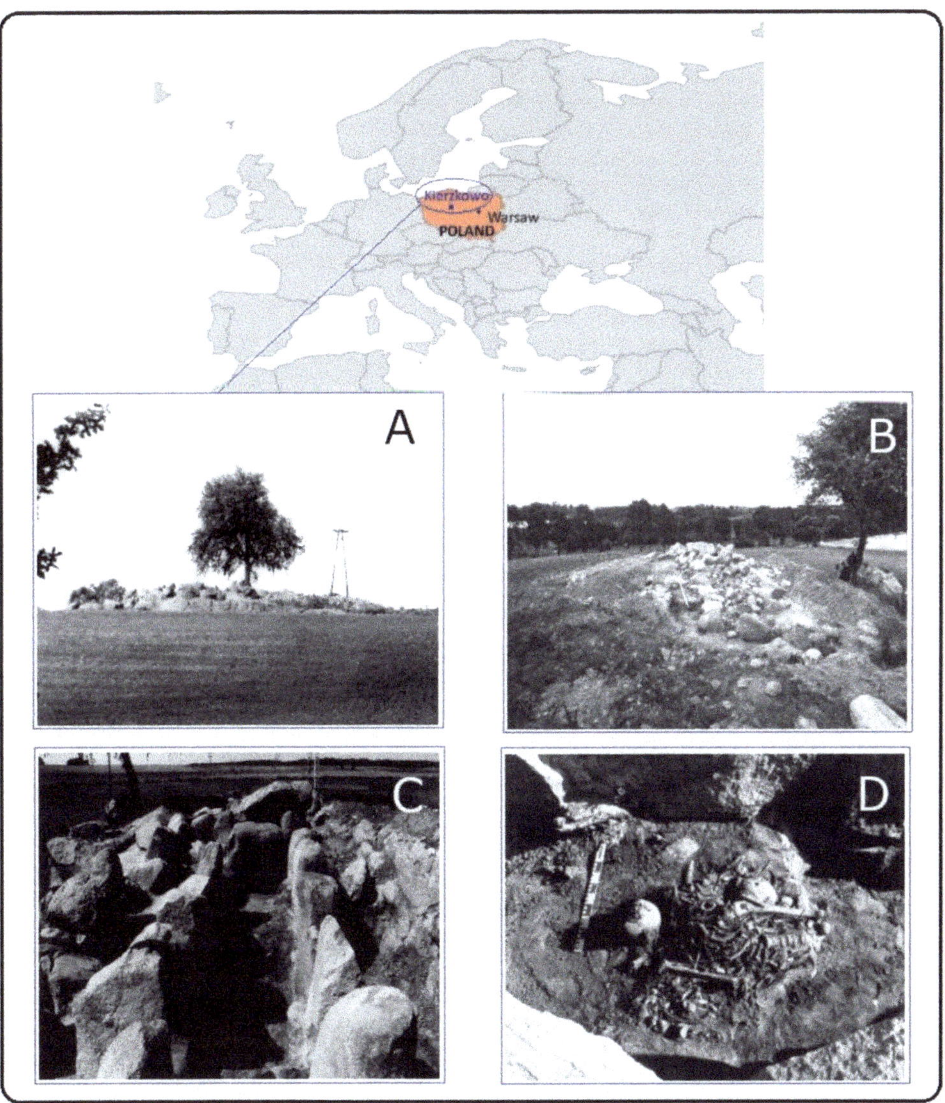

Figure 1. Geographic location and details of the site. Location of Kierzkowo site (Poland). (**A**) Megalithic barrow in Kierzkowo. (**B**) Megalithic barrow after removing the earth embankment. (**C**) Detail of the barrow: burial chamber. (**D**) Cluster of human Neolithic bones within the chamber. (**A**–**D**): photographed by Professor T. Wiślański; (**A**,**D**): fragments of photographs, adapted with permission from book [11]; (**B**,**C**): from the Archives of the Institute of Archaeology and Ethnology of the Polish Academy of Sciences.

Within the chamber, Neolithic human bones gathered into two large clusters and a smaller one (at metres 3 and 4 and 5, Figure 1D). Some bones were also located under the large stone dividing the chamber, and much fewer bones were found outside of the chamber. In the two large bone gatherings, remains were stratified into seven layers, while in other locations, bones were in superficial layers. In many instances, human bones were mixed with animal bones with signs of dismemberment [11]. The presence of domesticated animal remains testify that animal husbandry was the dominant form of economy. Individuals buried in Kierzkowo, indeed, being people of the GAC, belonged to the set of agriculture-based economies with a dominance of pastoralism. They were nomadic people with unstable settlement patterns. The megalithic tombs may have been important central points or boundaries of the territory used by the travelling human group [8,15–17]. Consumption of milk and animal meat is attested by the analysis of lipid residues in ceramic vessels from the barrow. Carbon ($\delta^{13}C$) and nitrogen ($\delta^{15}N$) isotopes were used to reconstruct the diet of the individuals from Kierzkowo. The $\delta^{15}N$ values in bone collagen were slightly higher in adult men than in adult women. This could indicate a slightly better access of men to valuable animal products, but this differentiation is very small and requires further research [11].

Most of the human remains found in Kierzkowo were highly fragmented and mixed, without anatomical order, placed in the barrow as secondary interment (Figures 1D and 2). Altogether, 428 fragments of various parts of the skeleton were recovered, 94.2% of them from the burial chamber. These fragments were studied for taphonomy and morphological traits. Type of bone, skeletal area, side of the body, size of a fragment, and its robustness surface markings were taken into account. Furthermore, the degree of bone deterioration, traces of plant roots and eventual activity of invertebrates, bone colour and its alterations, types of breaks and soil in which they were preserved were also considered. Wherever possible, sex and age at the death of individuals were estimated. Preliminary descriptions and anthropological analyses were made many years ago [18]. At present, all bone fragments were inspected again and revisions of diagnoses (such as estimations of sex and age) were performed using newer diagnostic methods.

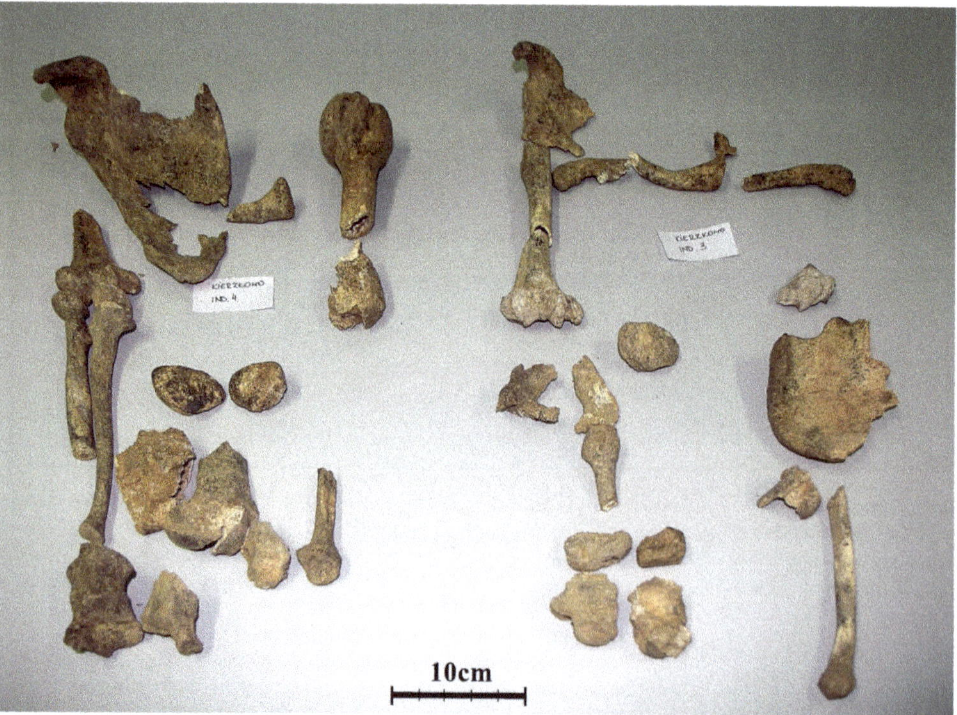

Figure 2. Fragments of bones of two individuals partly arranged in anatomical order. Bone fragments were tentatively attributed to individuals considering sex and age estimations, side, size, robustness, surface relief, colour, and state of preservation. Photo by Dr Andrzej Długoński.

Evaluating sex and age estimations, the origin from the left or right side of the skeleton, the number of paired bones in the skeleton, size, robustness, surface relief, and also colour and state of preservation, bone fragments were assembled into individual skeletons (Figure 2). It needs to be stressed that the application of only morphological criteria does not produce complete certainty about bone allocation to the same individual. The Ockham's razor principle was used for allocating bone fragments to individuals in order to not multiply numbers of individuals beyond the necessary minimum. Finally, we estimated that in the barrow there were at least 27 persons buried [11]. Both sexes and ages ranging from newborns to elders were represented.

2.2. The Analysed Samples

Seventeen bone remains from Kierzkowo were chosen for genetic analysis as they were attributed to the same number of putative different individuals (Table 1).

The primary reports of the molecular data obtained from these samples were previously published by [2,3], where population genetics methods were applied to these and other samples to investigate population dynamics during the Neolithic period in Europe. Complete or almost complete mtDNA consensus sequences were available in GenBank with accession numbers MF114211-MF114224. Sequence alignments of nuclear DNA (nuDNA) reads in BAM (Binary Sequence Alignment/Map) format were retrieved for eight individuals from the European Nucleotide Archive (ENA) under accession numbers ERS2040835-ERS2040840, ERS2040844, and ERS2040845.

In this paper, both mtDNA and nuDNA data were object to a new kind of analysis with a specific focus on the single community buried in the Kierzkowo site.

2.3. Mitochondrial Haplotype and Haplogroup Analysis

The available mtDNA consensus sequences were reanalysed to estimate haplogroups and phylogenetic relationships between haplotypes.

Different from our previous study [3], HaploGrep 2.0 [19] was used for haplogroup assignment based on PhyloTree Build 17 [20]. Polymorphisms according to the revised Cambridge reference sequence (rCRS) were annotated and showed in a graphical phylogenetic tree obtained by HaploGrep 2.0.

Misincorporation rates at read termini and average fragment length were calculated by MapDamage2.0 [21] to describe the damage pattern of the mtDNA molecules. Modern human contamination at mitochondrial level was estimated by ContamMix [22]. Default parameters were used for these analyses.

2.4. Sex Identification

Sex identification of all samples for which nuDNA data were available was performed using the guidelines suggested by Skoglund et al. [23] (v0.4). A python script was applied to the read alignments in BAM format to infer the molecular sex by comparing the number of alignments to the Y chromosome to the total number of alignments to X- and Y-chromosomes.

2.5. Y-Chromosome Haplogroup Determination

Y-chromosome haplogroup inference was attempted for male Neolithic samples using the software Yleaf v2.2 [24], which accepted read alignments in BAM format as input files. The prediction, performed with a minimum number of reads for each base (-r) set to 2 and a minimum quality for each read (-q) set to 30, was based on over 41,000 markers present in the International Society of Genetic Genealogy (ISOGG) Y-DNA Haplogroup Tree defining 5353 haplogroups [24].

2.6. Kinship Prediction

We first genotyped our read alignments in BAM format, after trimming 3 bases at read ends, by using bamUtil "trimBam" function (v1.0.14) [25], and SAMtools [26] "mpileup" function (v1.7), followed by pileupCaller from sequenceTools (v1.4.0.5, https://github.com/stschiff/sequenceTools, accessed on 1 August 2020) by randomly calling one allele per position considering the human genome as pseudo-haploid genome. Base alignment quality computation was disabled (-B) and minimum base (-q) and mapping quality (-Q) were set to 30 in the mpileup process, as recommended by the authors. We used GRCh37 as reference genome assembly and called the SNPs according to the 1240K panel [27–29]. Genotypes were obtained in Eigenstrat output format, subsequently converted in plink format using EIGENSOFT convertf (v7.2.1, https://github.com/DReichLab/EIG, accessed on 1 August 2020) and plink [30] (v1.9). Finally, kinship prediction was performed by READ [31], whose approach can successfully infer up to second degree relationships even with 0.1× shotgun coverage per genome for pairs of individuals. The analysis was carried out with default parameters, using the median across all pairwise differences for normalization.

In addition to READ, we also applied NgsRelate [32] (v2) to infer relatedness using genotype likelihoods instead of called genotypes. In this fashion, we compared kinship predictions obtained by two different approaches, useful for uncertain classifications due to low coverage. ANGSD [33] (v0.933) was used to calculate genotype likelihoods while doing SNP calling from the mapped reads with the GATK model [34], after trimming 3 bases from both read ends and requiring minimum base and mapping quality scores = 30. NgsRelate was run using European population allele frequencies, previously collected from the 1000 Genomes Project, phase 3 [35]. The tool calculated 11 summary statistics, including the coefficient of kinship theta [36], the KING-robust kinship statistics by Manichaikul et al. [37], and the "R0" and "R1" ratios described in Waples et al. [38], based on a two-dimensional site-frequency spectrum (2D-SFS) approach.

2.7. Analysis of Phenotypically Informative Markers

Variant calling of 41 SNPs from the HIrisPlex-S panel [39] related to eye, hair, and skin pigmentation was performed using two different approaches, allowing to get both pseudo-haploid and diploid calls on the same targeted sites. To the first aim, SAMtools mpileup and pileupCaller were used as described in the Section 2.6, while diploid calling was enabled by ATLAS [40] maximum likelihood genotype caller (task = call method = MLE) by specifying post-mortem damage and recalibration input files. These latter were previously obtained by ATLAS ("PMD" and "recal" functions, respectively) considering only bases with a minimum quality = 20 and using highly conserved positions among species defined by high GERP scores [41] as a training set to infer the recalibration parameters applied to our data. ATLAS allowed retrieving potential heterozygous variations, which could not be detected by the pseudo-haploid calling. We used the HIrisPlex-S [39,42–44] online web tool to predict eye, hair, and skin colour from DNA genotypes called by ATLAS. Moreover, 7 additional SNPs in *EDAR*, *MCM6*, *ABO*, and *RHCE* genes were also detected, being phenotypically informative markers. Annotations for each SNP identified in our samples were retrieved from SNPedia [45].

Table 1. The analysed specimens and summary of the results. Sample ID corresponds to the code attributed to the bone remains during the anthropological study, while Lab code corresponds to the laboratory label assigned during genotyping of nuclear genome. Note that Lab code is the same for sample 7.1 and 8.4, because the mitochondrial (mtDNA) and nuclear DNA (nuDNA) profiles indicated they belong to the same individual. Anatomical element, sex, and age determination obtained through anthropological methods [11,18] are indicated. Descriptive results for mtDNA and nuDNA analyses are reported. Direct radiocarbon dates are indicated where available. Samples 6.2, 6.3, 7.5, 8.8, and 8.9 were discarded from further analyses because of their low coverage or recent radiocarbon dating.

Sample ID	Lab Code nuDNA	Anatomical Element	Anthropological Sex and Age Estimation	mtDNA Molecules Average Length/ % CtoT at 5′ End	mtDNA Haplogroup	mtDNA Contamination Estimate (C.I.) %	SNPs Hit on Autosomes	Y-Chromosome Haplogroup	Molecular Sex Estimation	Conventional Radiocarbon Age	Calibrated Radiocarbon Result (95% Probability)
3.1	I2301	Left ulna	Male (50–X years)	67.6 bp/32	H1	6.56 (11.8–2.62)	-	-	-	-	-
3.4	I2403	Left femur	Male	63.8 bp/33	U5b2b1a1	0.08 (1.62–0.01)	154,174	I2a1b1	Male	4120 ± 30 BP (β-430,712)	Cal BC 2780 to 2575 (Cal BP 4730 to 4525)
5.1	I2433	Tooth/ mandible	Female (50–60 years)	61.5 bp/39	H28a	3.89 (5.21–2.86)	528,330	-	Female	-	-
5.3b	I2434	Right femur	Female (20–30 years)	65.2 bp/32	U5b1d1a	1.31 (2.27–0.71)	112,399	-	Female	-	-
6.1	I2435	Femur	Female (20 years or more, closer to 20 years)	59.8 bp/30	H28a	0.00 (0.001–0.00)	16,888	NA	Male	-	-
6.2	-	Right ulna	Female over 25	67.0 bp/11	J1c2	0.39 (0.58–0.01)	406,352	-	-	-	Historical, based on the association with 8.8 and 8.9
6.3	-	Mandible	Female	61.2 bp/-	-	-	-	-	-	-	-
7.1	I2407	Maxilla	Juvenile (about 14 years)	61.0 bp/38	H28a	2.30 (2.62–2.04)	36,339	NA	Male	-	-
7.5	-	Right humerus	Juvenile (14–19 years)	54.3 bp/-	-	-	-	-	-	-	-
7.6	I2440	Right humerus	Male? (20–50 years)	66.3 bp/23	H1b	0.61 (1.13–0.29)	189,493	I2	Male	-	-
8.1	I2801	Right femur	Child (2–3 years)	56.9 bp/40	H1b	0.80 (0.89–0.67)	-	-	-	-	-

Table 1. *Cont.*

Sample ID	Lab Code nuDNA	Anatomical Element	Anthropological Sex and Age Estimation	mtDNA Molecules Average Length/ % CtoT at 5′ End	mtDNA Haplogroup	mtDNA Contamination Estimate (C.I.) %	SNPs Hit on Autosomes	Y-Chromosome Haplogroup	Molecular Sex Estimation	Conventional Radiocarbon Age	Calibrated Radiocarbon Result (95% Probability)
8.2	I2405	Tibia	Child (7–8 years)	50.9 bp/33	W5	2.30 (5.47–0.39)	91,505	NA	Male	4460 ± 30 BP (β-430,713)	Cal BC 3335 to 3210 (Cal BP 5285 to 5160), Cal BC 3190 to 3150 (Cal BP 5140 to 5100), Cal BC 3140 to 3020 (Cal BP 5090 to 4970)
8.3	I2803	Right tibia	Child (14 years)	66.7 bp/38	H	0.94 (2.67–0.20)	-	-	-	-	-
8.4	I2407	Left femur	Child (10–14 years)	64.2 bp/32	H28a	0.34 (1.63–0.03)	36,339	NA	Male	4390 ± 30 BP (β-430,713)	Cal BC 3095 to 2915 (Cal BP 5045 to 4865)
8.5	I2441	Left pelvis	Newborn	64.0 bp/31	K1b1a1	3.86 (4.19–3.50)	510,373	I2a1b1a2b1~	Male		
8.8	-	Left femur	Child (infant)	63.7 bp/11	U5b2a2b1	0.50 (1.05–0.07)	298,820	-	-	210 ± 30 BP (β-430,715)	Cal AD 1645 to 1685 (Cal BP 305 to 265), Cal AD 1735 to 1805 (Cal BP 215 to 145), Cal AD 1930 to Post 1950 (Cal BP 20 to Post 0)
8.9	-	Right femur	Child (2–3 years)	62.1 bp/12	U3b3	0.22 (2.71–0.02)	419,659	-	-	130 ± 30 BP (β-430,716)	Cal AD 1670 to 1780 (Cal BP 280 to 170), Cal AD 1800 to Post 1950 (Cal BP 150 to Post 0)

3. Results and Discussion

3.1. Mitochondrial Haplotype and Haplogroup Analysis

Our molecular study of the human skeletal remains from Kierzkowo first focused on the analysis of mitochondrial DNA (mtDNA) genomes obtained through a capture enrichment approach associated with next generation sequencing [3]. This strategy allowed us to obtain 15 complete or almost complete mtDNA profiles out of 17 samples. Polymorphisms according to rCRS are annotated in Table S1 and showed in a phylogenetic tree including information about haplogroup attribution (Figure S1). The C > T misincorporation percentages at 5′ termini and average fragment length as well as estimates of modern human contamination calculated for the mtDNA molecules are indicated in Table 1.

As expected, the reanalysis of haplogroup attributions based on a newer version of PhyloTree and HaploGrep agrees with previously published data [3], except for some samples now attributed to specific sub-lineages generally with a higher resolution (H28a instead of H28, U5b2b1a1 instead of U5b2b1, U5b1d1a instead of U5b1d1, U3b3 instead of U3b, and H1 instead of H1b for sample 3.1). Haplogroup H was the most frequently predicted and represented more than half of the mtDNA variation in our dataset with eight samples carrying three different lineages (H, H1 and H28a). Haplogroup U characterises four samples, three belonging to U5 lineages and one to U3b. The remaining three samples belong to haplogroups J, K, and W. At the haplotype level, despite some missing positions, four samples (corresponding to sample IDs 7.1, 8.4, 6.1, and 5.1) share the same mutation motif, which is attributed to the H28a haplogroup. This haplotype sharing could be explained (i) by a possible maternal relationship, more or less close in time, or (ii) by the presence of two or more samples belonging to the same individual (as shown below, further analyses at the nuclear genome level were necessary to support one of the two possible hypotheses).

The evaluation of damage patterns described through average fragment length and the C > T misincorporation percentage at 5′ termini showed a certain degree of molecular degradation for all the samples. Even if fragment length cannot be used as an antiquity marker (degradation of the DNA to an average size lower than 100 bp occurs rapidly after the death of the organism, probably due to autolytic processes [46,47]), the average length of our samples was estimated to range from 50.6 to 67.7 bp, in line with degradation processes.

Moreover, a different level of cytosine deamination at read ends was observed for the samples. Most of them showed percentages of C > T at 5′ ranging from 23% to 39% (values compatible with the Neolithic chronology of the site [46]), while three samples (6.2, 8.8, and 8.9) were characterised by lower percentages of molecules affected by misincorporation ranging from 11% to 12%.

As is known, opposite to fragment length, the frequency of misincorporations at the ends of the molecules tends to increase over time and can be used to distinguish modern from ancient DNA (aDNA) [46]. Therefore, the presence of a lower percentage of misincorporation for samples 6.2, 8.8, and 8.9 could be a signal of possible modern contamination of the samples or indicate that the three individuals did not belong to the Neolithic period.

The estimates obtained by the evaluation of ContamMix [22] allowed us to exclude a relevant contamination by modern human DNA for these samples. Referring to archaeological data, these three samples turned out to belong to a skeletal assemblage located outside the principal burial chamber of the tomb.

For these reasons, as described below, further investigation was required to explain the different damage patterns that characterise these samples.

3.2. Analysis of Nuclear Genome Data

Nuclear genome data were obtained through a target enrichment approach based on using 1240 K informative SNPs as a target [2]. The population genetics analysis showed the three individuals characterised by lower misincorporation percentages as diverging from the overall genetic variability associated with the other samples from Kierzkowo, providing further clues about a possible different provenance of these samples [2]. Radiocarbon dating was, therefore, applied to some of the bones (both with lower and higher values of misincorporation percentages) in order to directly check their chronology. Dating results confirmed the bones from the tomb chamber as belonging to the Globular Amphorae culture (GAC) period: 2870–2575 cal BCE (β-430,712, sample 3.4), 3260–3095 cal BCE (β-430,713, sample 8.2), and 3015 cal BCE (β-430,714, sample 8.4) [2]. Results obtained for the bones found outside the burial chamber proved, instead, that 6.2, 8.8, and 8.9 samples represented a recent intrusive burial (8.8 and 8.9 dated to 210 ± 30 BP and 130 ± 30 BP, respectively, β-430,716 and β-430,715 [2]) of an adult woman and two children (one infant and a 2–3 year-old child), who were not maternally related, according to the mtDNA profiles. As a result, archaeologists suggested that they may have been people who—for some reason—could not be buried in the "consecrated ground" or that they were victims of one of the epidemics that ravaged these lands, such as the plague in the early 18th century or the cholera epidemic in the 19th century. In fact, at some distance from Kierzkowo, a mass grave of the victims of the pestilence from the mid-19th century was discovered. In addition, several victims of the cholera epidemic were also buried in the 19th century in the megalithic tomb of the GAC in Złotowo, 10 km from Kierzkowo [11].

If data obtained from the study of mtDNA genomes first provided important information about possible relations among samples and their chronology, only with the analysis of nuclear DNA (nuDNA) data was it possible to clarify the clues arising from the mtDNA analysis and to obtain thorough attributions at the individual level. Nuclear genome analysis, indeed, allowed us to clarify the relationships among the four samples sharing the same mitochondrial H28a haplotype. By evaluating the mitochondrial and nuclear DNA data, it was possible to provide a proper estimate of the minimum number of individuals (MNI), which otherwise would not have been allowed due to the complex taphonomic situation of bone remains in the burial.

In addition, the comparison of the obtained nuDNA genotypes allowed us to identify and verify that samples 7.1 and 8.4 belonged to the same individual and aid the anthropological study in the correct site reconstruction since the two samples (a maxillary and femur fragment) had been previously tentatively attributed to different individuals because of surface appearance and state of preservation.

3.2.1. Sex Identification

Nuclear genome data were obtained for eight individuals, of which six were identified as male and two as female (Table 2). The comparison of genetic and anthropological data highlighted a perfect match of results for all samples analysed, except sample 6.1 for which the genetic determination gave a result different from the anthropological estimation. Furthermore, it generally allowed us to obtain a determination in case of absence of diagnostic morphological markers, i.e., for highly fragmented remains and infant and juvenile individuals. As a result, genetic sex determination allowed us to improve the attributions previously made through the anthropological approach. Even in this case, the fragmentation and preservation level of skeletal material, as well as the presence of infant and juvenile individuals, represented a strong limit for proper sex determination via the anthropological approach.

3.2.2. Y-Chromosome Haplogroup Determination

Three out of six male individuals yielded enough valid markers to allow a haplogroup determination with the approach used by Y-leaf 2.2 [24]. All three samples (3.4, 7.6, 8.5) belonged to haplogroup I2, with different levels of definition at the sub-haplogroup level (Table 3). As expected, the more valid markers were available per sample, the deeper attribution at the sub-haplogroup level was possible. Due to the limited number of shared covered positions between samples and their low coverage, it was not possible to perform a detailed comparison between individuals at the haplotype level. For this reason, we have been able to highlight the presence of the same haplogroup in more individuals but not go deeper into the estimate of haplotype sharing and possible kin relationships based on this uniparental marker.

Table 2. Sex reconstruction for the Neolithic bones. Nseq: total number of sequences; NchrY + NchrX: total number of alignments to both sex chromosomes; NchrY: number of alignments to the Y chromosome; R_y: ratio between NchrY and NchrY + NchrX; SE: R_y standard error; 95% CI: 95% confidence interval for R_y.

Sample ID	Nseqs	NchrY + NchrX	NchrY	R_y	SE	95% CI	Assignment
3.4	690,382	18,622	5780	0.3104	0.0034	0.3037–0.317	XY
5.1	2,074,197	74,852	334	0.0045	0.0002	0.004–0.0049	XX
5.3	267,061	9152	59	0.0064	0.0008	0.0048–0.0081	XX
6.1	37,173	979	284	0.2901	0.0145	0.2617–0.3185	XY
7.1/8.4	19,211	522	155	0.2969	0.02	0.2577–0.3361	XY
7.6	382,899	10,099	3389	0.3356	0.0047	0.3264–0.3448	XY
8.2	195,536	5291	1663	0.3143	0.0064	0.3018–0.3268	XY
8.5	1,783,600	48,164	14,351	0.298	0.0021	0.2939–0.302	XY

3.2.3. Kinship Analysis

Kinship predictions by READ software [31] allowed us to identify the couples 6.1–7.6, 7.1/8.4–5.1, and 5.1–6.1 as possible parent–offspring or full-siblings, while 7.1/8.4–6.1 and 7.1/8.4–7.6 were classified as second-degree relatives (grandparent–grandchild, avuncular or half-siblings). However, the standard error of the average proportion of non-matching alleles (P0 values) calculated by READ [31] expressed very high uncertainty of the classification for the couple 7.1/8.4–6.1 (Figure 3), probably due to a reduced number of comparable genomic positions. To a lesser extent, the categorisation of most couples showed uncertainty, frequently spanning more than one class; thus, only the first-degree relationships involving sample 5.1 may be considered unambiguously predicted (Figure 3). We also performed a genotype likelihood-based analysis of kinship using NgsRelate [32] to compare kinship predictions obtained using different approaches (see Methods). Both methods led to quite similar outcomes and the full results are shown in Supplementary Table S2. First degree relationships involving 6.1 were confirmed and further distinguished using the R0 ratios described in Waples et al. [38] into parent–offspring or sibling relationships [38], resulting in a family unit composed of 5.1 as mother, 7.6 as father and 6.1 as son. A maternal relationship between mother and son was also confirmed by the same mtDNA profile belonging to haplogroup H28a. Theta values suggested a first degree of kinship for the couple 7.1/8.4–6.1; however, the prediction remained unreliable since only 189 genomic sites were considered in the analysis. A first-degree relationship was also estimated for the couple 7.1/8.4–7.6, differing from the READ prediction (second-degree). Low coverage, low degree of preservation of aDNA in these samples and possible inbreeding leading to a discrepancy from expected values of identity-by-descent (IBD), as shown for samples 7.1/8.4 and 6.1 (Table S2), could explain the difficulties in inferring kinship. Individual 7.1/8.4 could be, indeed, brother or uncle of 6.1, as well as brother of 5.1 as indicated also by mtDNA analysis; furthermore, he could be involved in a possible second-degree relationship with 7.6. Both biological and methodological reasons could complicate pedigrees, e.g., the previous generation showing some degree of relatedness or inaccurate population allele frequencies due to population structure in the data. The different kinship degree predicted by the two software programs could be explained by another possible reconstruction where individual 7.1/8.4 could be the half-brother of 6.1, (both sons of 5.1) with a possible kinship relation between their fathers.

Table 3. Y-chromosome haplogroup determination. Hg: final haplogroup predicted using ISOGG nomenclature. Hg_Marker: final haplogroup predicted using marker nomenclature. Total_Reads: total number of reads. Valid_Markers: number of markers used for the haplogroups prediction. QC_score: overall quality score that is the factor of the three scores QC-1, QC-2, and QC-3; QC-1: score that indicates whether the predicted haplogroup follows the expected backbone of the haplogroup tree structure; QC-2: score that indicates whether equivalent markers to the final haplogroup prediction were found in the ancestral state; QC-3: score that indicates whether the predicted haplogroup follows the expected within haplogroup tree structure.

Sample ID/Lab Code	Hg	Hg _Marker	Total_Reads	Valid_Markers	QC-Score	QC-1	QC-2	QC-3
3.4	I2a1b1	I-FGC3552/etc* (xY10631,Y33740,Y7225, A11214, Y5728,Y11154,Y16445,Y5224,Y6098)	527,527	526	1.0	1.0	1.0	1.0
6.1	NA	NA	41,862	11	0	0	0	0
7.1/8.4	NA	NA	21,520	7	0	0	0	0
7.6	I2	I-CTS8444* (xPF4049,Y22356,Y24165,Y7822, Y5728,S17264,Y68424,Z45444)	307,275	274	1.0	1.0	1.0	1.0
8.2	NA	NA	175,789	48	0	0	0	0
8.5	I2a1b1a2b1~	I-FGC3587*(xFGC3570)	1,056,530	2025	1.0	1.0	1.0	1.0

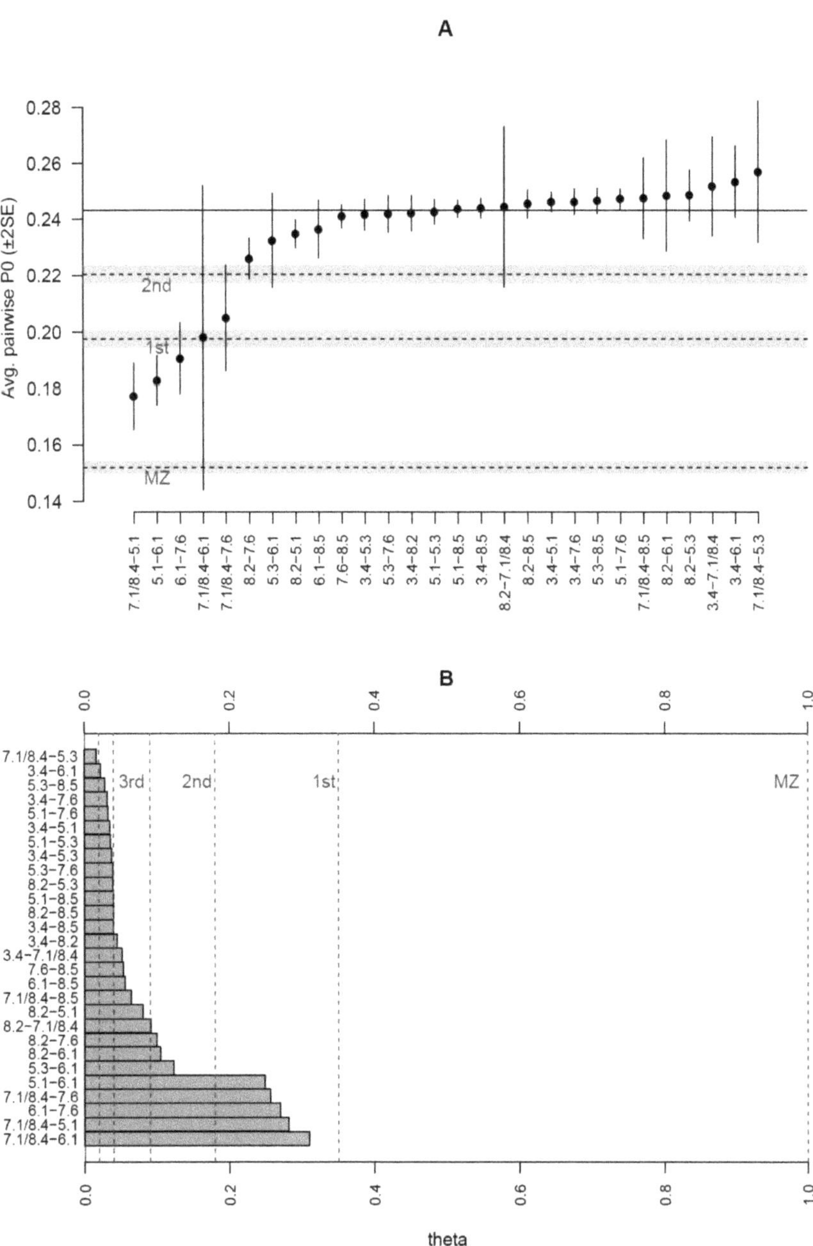

Figure 3. Kinship predictions by READ and NgsRelate software. (**A**) The average proportion of non-matching alleles (P0 values) with two standard errors calculated by READ is shown for each pair of samples. The solid horizontal line indicates the median value used for normalization. Dashed lines show the cut-offs calculated by READ to classify individuals as identical twins (MZ), first-degree, or second-degree related. The grey areas around dashed lines indicate 95% confidence intervals for the cut-offs. (**B**) Theta values for each pair of samples were calculated by NgsRelate. Dashed lines define ranges of expected theta values for each kinship category. We considered predictions of kinship by NgsRelate up to the third degree.

Additional second-degree relationships could be inferred from the theta values involving 5.3 and 8.2 (although they were not confirmed by READ), as well as potential third-degree relationships for nine couples of samples. Moreover, the relationship for all the remaining pairs seemed to be more distant than that of third-degree, but it could be overestimated by NgsRelate, as shown for the old version of the tool [48]; thus, we considered these couples as unrelated. The KING statistic is frequently the most discordant among all the estimated predictors: it probably underestimates kinships, as already demonstrated [49].

As a result of kinship analysis, we were able to construct a reasonable possible pedigree tree including all four abovementioned samples by admitting the absence of some members of a hypothetical extended family among the remains (Figure 4), with the nuclear family composed of 5.1 (mother), 7.6 (father), and 6.1 (son), and considering 7.1/8.4–6.1 as half-siblings and 7.1/8.4–7.6 as nephew/uncle.

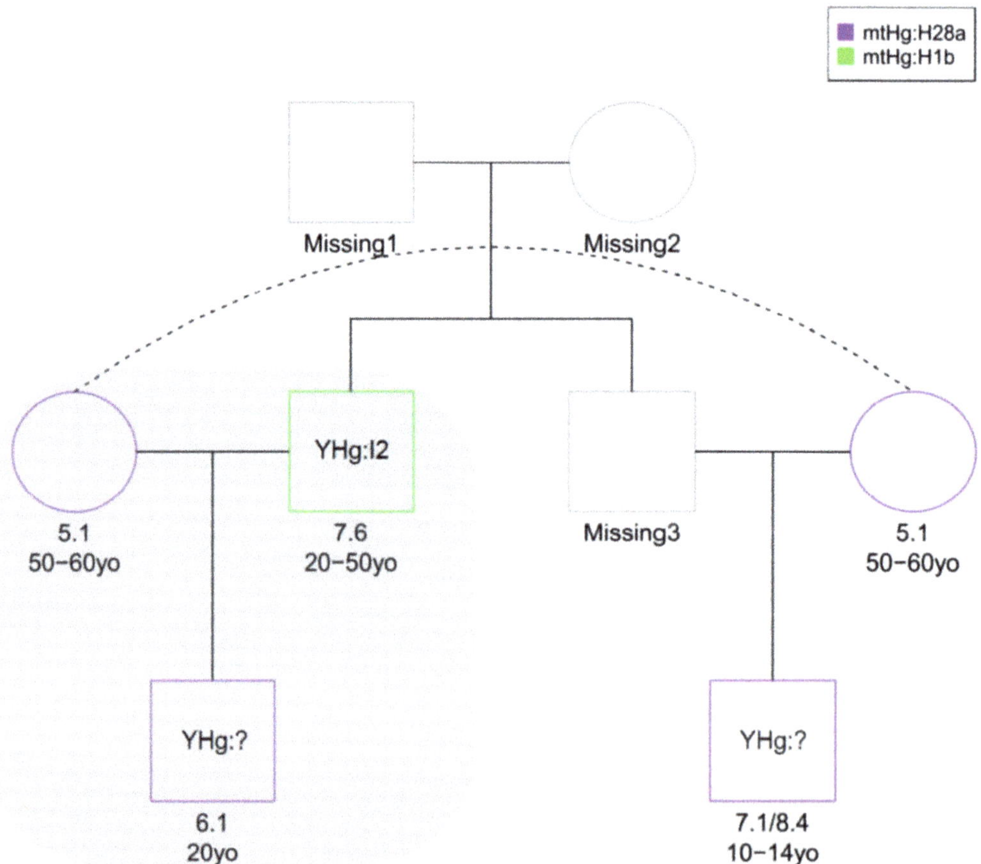

Figure 4. Pedigree reconstructed for the Kierzkowo family. A possible pedigree was inferred from genome-wide estimates of pairwise identity-by-descent (IBD). The nuclear family in the grey area was confirmed by both READ and NgsRelate tools. For each individual, age (yo: years old) estimated through morphological analysis is indicated. Y chromosome (YHg) haplogroup is described where available. Mitochondrial haplogroups (mtHg) are highlighted in different colours. Mitochondrial haplotype sharing supports the estimates of kinship relations between mother and sons based on genomic data (see Figure S1 and Table S1). The pedigree was plotted using the kinship2 R package [50].

In an attempt to combine molecular with anthropological data, information about age classes estimated through morphological analysis could help to discriminate between different hypotheses about kinship relations, but we have to consider the uncertainty of the age estimates due to high fragmentation and degradation of skeletal remains and the possibility that individuals could have been dead and buried in different epochs.

Even if the obtained data did not allow a sound detailed reconstruction of the possible pedigrees, we can suppose the presence of at least a nuclear family and of different degrees of biological relationships between several couples of individuals buried in the Kierzkowo barrow. Even in this case, mtDNA data provided a first indication about possible relationships among individuals, but nuDNA information was necessary to discriminate between different possible reconstructions.

3.2.4. Phenotypes

Because of low or no coverage, we were able to partially detect the selected phenotypically informative markers in our samples. Indeed, 32 of 41 HIrisPlex-S SNPs were not totally covered in some samples, presumably due to DNA fragmentation and because the remaining nine SNPs were not part of the 1240 K capture array. Generally, diploid genotyping (see Methods) called 35 out of the 48 analysed genomic positions (41 HIrisPlex-S + 7 additional markers), including three variants in a heterozygous state and 28 homozygous alleles (either reference or variant) confirmed by pseudo-haploid calling (Table S3). Samples 5.1, 6.1, and 8.5 probably had blue eyes, as predicted by the HIrisPlex-S tool [39], and confirmed by data reported in SNPedia 6 [45]. The same eye colour could be supposed for 3.4, 5.3, and 7.6 (SNPedia, Table S3). Hair colour was predicted for 5.1 (dark brown) and 8.5 (blonde), while 6.1 was probably light-haired. Red hair could be excluded as a phenotypic trait in more than half of the samples (3.4, 8.2, 5.3, in addition to 5.1 and 8.5). In terms of skin colour, pale skin was predicted for 8.5 and 5.1, which was also confirmed by SNPedia. The absence of freckles in 8.5 and light skin colour in 5.3 can be additionally supposed. Annotations by SNPedia are shown in Supplementary Table S3; probabilities calculated by the HIrisPlex-S tool are reported in Table 4 and Supplementary Table S4.

We also investigated *AB0* and *RHD* genes. Blood type 0 was inferred for 8.5, a classic OO homozygote like the 42% of Caucasians [45], while we could just exclude type B as the blood group for 3.4, 5.1, and 7.6, whose specific blood group could not be further determined due to a lack of coverage of the genomic position related to rs8176719 in the *ABO* gene. Similarly, Rhesus blood groups could not be properly inferred in our data since rs676785 and rs609320 in the *RHCE* gene (which determine c and e antigens, respectively) are not included in the 1240K SNPs panel. However, the *RHD* gene was partially covered in all the samples. Thus, its full deletion, which is mainly associated with the Rh-phenotype in Europeans [51], can be excluded.

Another trait we investigated is lactase persistence, since consumption of milk is attested in Kierzkowo by the presence of milk fats in ceramic vessels found in the burial and reconstruction of the diet through isotope analysis of human bone remains. Lactase is, indeed, the enzyme responsible for the digestion of the milk sugar lactose. Its production usually decreases after the weaning phase, except for some individuals who continue to produce lactase throughout adulthood. A mutation at the SNP rs4988235 (−13,910*T) in the *MCM6* gene, supporting the lactase persistence haplotype, explains the phenotype in European populations [52]. We were able to describe this SNP only in four individuals (5.1, 7.6, 8.2, and 8.5) from Kierzkowo. Interestingly, they all were homozygous for the ancestral allele, suggesting they could have become lactose intolerant in adulthood, probably with no effect on 8.5, who was a newborn. It is a common idea that lactase persistence coevolved with the cultural adaptation of dairying. Interestingly, the date estimates for the rise in frequency of the −13,910*T mutation (2188 and 20,650 years ago [53]) bracket archaeological dates for the spread of domestic animals and dairying into Europe, pointing at a probably strong positive selection linked to the cultural traits of animal domestication and adult milk consumption [53–56]. However, aDNA studies have shown that the −13,910*T allele was very rare or absent in early Neolithic central Europeans, which lets us suppose that lactase persistence and dairying coevolution began around 7500 years ago, probably among people of the Linearbandkeramik culture between the central Balkans and central Europe [57]. According to our data, people from Kierzkowo were lactose intolerant in adulthood, even though their culture was characterised by the importance of dairying. We have to consider that an inter-individual variation of the amount of lactose tolerated by lactase non-persistent people (probably as a result of variation in the composition of the gut flora) is known and that some intolerant individuals can consume lactose-containing products without any obvious ill effects. In particular, fermented dairy products (i.e., yoghurt or cheese) are usually well tolerated by non-persistent individuals since they contain less lactose [58].

3.3. Comparison between GAC Sites

A comparison of our results can be made with information available from another burial site attributed to the GAC in Poland, for which genome analysis was performed on human bone remains: the mass grave of Koszyce (Małopolska province) [59]. In this case, 15 individuals including men, women and children, were buried at the same time, next and on top of each other, after a violent death. According to the genomic data, the buried individuals belonged to a large extended family connected by several first- and second-degree relationships. In particular, four nuclear families were represented. The presence of unrelated females and related males would indicate a dominant form of patrilineal social organisation. The two burial contexts are quite different (a mass grave in Koszyce and a typical GAC burial ritual in Kierzkowo), as well as the experimental strategy applied due to a different level of sample preservation (whole genome sequencing for Koszyce samples and target enrichment of selected SNPs for Kierzkowo). However, similarly genetic and reproductive relationships were probably the basis of social relationships in GAC communities. In Kierzkowo as well as in Koszyce, the presence of different mtDNA lineages can support the hypothesis of a community not based on the matrilocal residence system. Unlike the Koszyce samples, the resolution of our data on the Y-chromosome did not allow us to confirm the presence of a unique patrilineal lineage, although we cannot exclude it either, since data for Kierzkowo show the same haplogroup for all the available samples. In particular, Y-chromosome variability is represented in both sites only by the I2 haplogroup, indicating a low genetic diversity in the GAC population according to this marker, which could support the hypothesis of a society based on patriarchy.

This hypothesis could find support from archaeological data: the economic base of the GAC group from Kierzkowo was agriculture and animal husbandry. In particular, the latter played a special role, as evidenced by the numerous skeletal remains of farm animals and the milk fat residues retrieved in ceramic vessels found in the burial, which suggest consumption of milk and meat. Preliminary isotope analysis on the human skeletal remains from Kierzkowo confirms the presence of these foods in the diet and seems to indicate better access to valuable animal products for men than for women [11]. Furthermore, most pastoralist societies are patriarchal due to the male role in animal husbandry. A particular role of men in GAC society is also sometimes demonstrated by the equipment of the graves, which is differentiated by gender [8,15,60–63].

Table 4. HIrisPlex-S annotations and probabilities. Prediction probabilities calculated by the HIrisPlex-S software are shown only for the three samples with available data. Complete results are provided in Supplementary Table S4. In terms of eye colour, the predicted phenotype (blue) was defined by the highest p-values. In terms of hair colour, the highest p-value approach is used in combination with a stepwise model as described in [43]. In terms of skin colour, the highest p-value approach is used in combination with the second highest probability value [39].

Sample ID	Blue Eye	Intermediate Eye	Brown Eye	Blond Hair	Brown Hair	Red Hair	Black Hair	Light Hair	Dark Hair	Very Pale Skin	Pale Skin	Intermediate Skin	Dark Skin	Dark to Black Skin
5.1	0.898	0.063	0.039	0.371	0.467	0.008	0.154	0.779	0.221	0.081	0.539	0.380	0.000	0.000
6.1	0.912	0.057	0.031	NA	NA	NA	NA	0.934	0.066	NA	NA	NA	NA	NA
8.5	0.914	0.055	0.031	0.651	0.216	0.113	0.020	0.953	0.047	0.123	0.656	0.220	0.000	0.000

4. Conclusions

Due to the (quite recent) introduction of next generation sequencing methodologies in the aDNA field, particular emphasis has been placed on archaeobiological research (archaeobotany, zooarchaeology, anthropology) as a way to gain insight into the environmental conditions and the social organisation of past populations and provide information at the individual level.

The study of the human bone remains found in the Kierzkowo Megalithic barrow allowed us to add important details to the description of this Neolithic community associated with the Globular Amphorae culture, and to the reconstruction of the use of the site through time. Indeed, the evaluation of nucleotide misincorporation patterns of aDNA (molecular degradation) suggested the presence—confirmed through radiocarbon C14 dating—of a recent intrusion into the burial that happened in the last century, when the remains of a woman and two little children were deposed inside the perimeter of the Neolithic burial. Together with the more recent remains, two female and six male Neolithic individuals were found in the tomb and identified. Despite the taphonomic condition of the site, characterised by a secondary deposition of fragmented and mixed bone remains, kin relationships and phenotypic traits of the buried individuals were also successfully estimated via aDNA analysis. Most of them were related by first-, second-, and third-degrees of kinship, and in particular, we located a nuclear family, one of the oldest attested by genetic data.

Considering the nuDNA data in general, the methodological approach used presented some limitations in terms of information recovered, as results are restricted to a relatively small and pre-selected set of genomic markers sequenced after a target enrichment. A whole-genome sequencing with high coverage would represent, indeed, the best strategy to obtain accurate individual profiling. However, producing whole-genome data is suitable for samples with levels of endogenous DNA content higher than that in Kierzkowo bone fragments (average 0.19%, with a minimum value of 0.007% and a maximum value of 0.66% for Neolithic individuals). More suitable samples than those used would have been represented by the petrous part of the temporal bone or by teeth, but they were not available for this study. Thus, to the best of our knowledge, the methodology chosen was the most suitable.

In conclusion, thanks to the development of the new sequencing technology and the choice of the best experimental strategy according to the level of preservation and the type of sample available, this study highlighted a new opportunity for ancient genomic studies.

Despite its limitations, even a target enrichment approach originally conceived for investigating large-scale populations' structure and relationships can provide interesting information at the individual level that can be exploited for a fine reconstruction of singular archaeological contexts. Through a multidisciplinary approach that involves archaeologists, archaeobotanists, zooarchaeologists, and biological and molecular anthropologists, a detailed bioarchaeological reconstruction of ancient communities and their members can be supplied. From this perspective, molecular data represent an important informative source to support biological anthropological activity, answering specific questions and reconstructing individual profiles as well as biological and social relationships. The results presented in this study, together with possible further genetic data obtained for other GAC sites, can provide an important contribution that is useful for supporting archaeological and anthropological research on the social organisation and funerary rituals that characterised this culture.

Supplementary Materials: The following are available online at https://www.mdpi.com/article/10.3390/genes12060910/s1, Table S1: Polymorphisms according to rCRS and Mitochondrial Haplogroup determination, Table S2: Kinship analysis by READ and NgsRelate, Table S3: Annotation of phenotypically informative SNPs detected by ATLAS in the dataset, Table S4: Eye, hair, and skin colour prediction according to HIrisPlex-S on SNPs detected by ATLAS, Figure S1: Phylogenetic tree and haplogroup determination for fifteen mitogenomes from Kierzkowo. Polymorphisms are indicated according to rCRS reference sequence. Haplogroup assignment is based on PhyloTree Build 17. For the sake of simplicity, sample IDs were used to indicate samples.

Author Contributions: Conceptualization, S.V. and E.P.; methodology, S.V., M.A.D., E.P.; formal analysis, S.V., M.A.D., C.C.; investigation, S.V., M.A.D., C.C., A.B., E.P.; resources, S.V., M.L., D.C.; data curation: S.V., M.A.D., E.P.; visualization, S.V., M.A.D., C.C., E.P.; writing—original draft preparation, S.V., M.A.D., A.B., E.P.; writing—review and editing, S.V., M.A.D., C.C., A.B., M.L., D.C., E.P. All authors have read and agreed to the published version of the manuscript.

Funding: This research was funded by the Italian Ministry of Education, University and Research (PRIN2017 grant no. 20177PJ9XF) and by the University of Florence—Bando di Ateneo per i progetti competitivi per Ricercatori a Tempo Determinato (RTD) 2019-2020 (Prot. n. 197432).

Institutional Review Board Statement: Ethical review and approval was not required for the study on human participants in accordance with the local legislation and institutional requirements. Written informed consent for participation was not required for this study in accordance with the national legislation and the institutional requirements.

Informed Consent Statement: Not applicable.

Data Availability Statement: The data analysed in this study are openly available in the European Nucleotide Archive (ENA) under accession numbers ERS2040835-ERS2040840, ERS2040844, and ERS2040845.

Conflicts of Interest: The authors declare no conflict of interest.

References

1. Nielsen, R.; Akey, J.M.; Jakobsson, M.; Pritchard, J.K.; Tishkoff, S.; Willerslev, R.N.E. Tracing the peopling of the world through genomics. *Nature* **2017**, *541*, 302–310. [CrossRef]
2. Mathieson, I.; Alpaslan-Roodenberg, S.; Posth, C.; Szecsenyi-Nagy, A.; Rohland, N.; Mallick, S.; Olalde, I.; Broomandkhoshbacht, N.; Candilio, F.; Cheronet, O.; et al. The genomic history of southeastern Europe. *Nature* **2018**, *555*, 197–203. [CrossRef]
3. Tassi, F.; Vai, S.; Ghirotto, S.; Lari, M.; Modi, A.; Pilli, E.; Brunelli, A.; Susca, R.R.; Budnik, A.; Labuda, D.; et al. Genome diversity in the Neolithic Globular Amphorae culture and the spread of Indo-European languages. *Proc. R. Soc. B Biol. Sci.* **2017**, *284*, 20171540. [CrossRef]
4. Veeramah, K.R. The importance of fine-scale studies for integrating paleogenomics and archaeology. *Curr. Opin. Genet. Dev.* **2018**, *53*, 83–89. [CrossRef]
5. Johnson, K.M.; Paul, K.S. Bioarchaeology and kinship: integrating theory, social relatedness, and biology in ancient family research. *J. Archaeol. Res.* **2016**, *24*, 75–123. [CrossRef]
6. Vai, S.; Amorim, C.E.G.; Lari, M.; Caramelli, D. Kinship determination in archeological contexts through DNA analysis. *Front. Ecol. Evol.* **2020**, *8*, 83. [CrossRef]
7. Fortes, G.G.; Speller, C.F.; Hofreiter, M.; King, T.E. Phenotypes from ancient DNA: Approaches, insights and prospects. *Bioessays* **2013**, *35*, 690–695. [CrossRef]
8. Hensel, W. *Polska Starożytna*; Zakład Narodowy im. Ossolińskich: Wrocław, Poland, 1980. (In Polish)
9. Wiślański, T. *Kultura Amfor Kulistych w Polsce Północno-Zachodniej*; Zakład Narodowy im. Ossolińskich: Wrocław, Poland, 1966. (In Polish)
10. Wiślański, T. Dalszy rozwój ludów neolitycznych. Plemiona kultury amfor kulistych. In *Prahistoria Ziem Polskich*; Hensel, W., Wiślański, T., Eds.; Zakład Narodowy im. Ossolińskich: Wrocław, Poland, 1979; Volume 2, pp. 261–299. (In Polish)
11. Nowaczyk, S.; Pospieszny, Ł.; Sobkowiak-Tabaka, I. A Megalityczny grobowiec kultury amfor kulistych z Kierzkowa: Milczący świadek kultu przodków z epoki kamienia. *Biskup. Archaeol. Work.* **2017**, *12*, 375p. (In Polish)
12. Bakker, J.A. *The Ditch Hunebedden. Megalithic Tombs of the Funnel Beaker Culture*; International Monographs in Prehistory: Ann Arbor, MI, USA, 1992.
13. Wiślański, T. *Kierzkowo, gm. Jadowniki, woj. Bydgoskie. Stanowisko. Sprawozdanie z 1982 Roku*; Zakład Archeologii Wielkopolski IHKM PAN w Poznaniu: Poznań, Poland, 1982. (In Polish)
14. Wiślański, T. *Kierzkowo, stan. 1, gm. Jadowniki k. Żnina. Sprawozdanie z 1982 roku*; Zakład Archeologii Wielkopolski IHKM PAN w Poznaniu: Poznań, Poland, 1982. (In Polish)
15. Ciesielska, A. *Społeczeństwa Europy Pradziejowej*; Wydawnictwo Naukowe UAM: Poznań, Poland, 2011. (In Polish)
16. Gąssowski, J. *Kultura Pradziejowa na Ziemiach Polski*; Państwowe Wydawnictwo Naukowe: Warsaw, Poland, 1985. (In Polish)
17. Godłowski, K.; Kozłowski, J.K. *Historia Starożytna Ziem Polskich*; Państwowe Wydawnictwo Naukowe: Warsaw, Poland, 1976. (In Polish)
18. Budnik, A.; Wrzesiński, J. Kierzkowo—Między inhumacją a ciałopaleniem. In *Popiół i Kość*; Wrzesiński, J., Ed.; Muzeum Ślężańskie im. S. Dunajewskiego w Sobótce: Sobótka, Poland; "AKME" Zdzisław Wiśniewski: Wrocław, Poland, 2002; pp. 125–145. (In Polish)

19. Weissensteiner, H.; Pacher, D.; Kloss-Brandstatter, A.; Forer, L.; Specht, G.; Bandelt, H.J.; Kronenberg, F.; Salas, A.; Schonherr, S. HaploGrep 2: Mitochondrial haplogroup classification in the era of high-throughput sequencing. *Nucleic Acids Res.* **2016**, *44*, W58–W63. [CrossRef] [PubMed]
20. Van Oven, M.; Kayser, M. Updated comprehensive phylogenetic tree of global human mitochondrial DNA variation. *Hum. Mutat.* **2009**, *30*, E386–E394. [CrossRef] [PubMed]
21. Jónsson, H.; Ginolhac, A.; Schubert, M.; Johnson, P.L.F.; Orlando, L. mapDamage2.0: Fast approximate Bayesian estimates of ancient DNA damage parameters. *Bioinformatics* **2013**, *29*, 1682–1684. [CrossRef] [PubMed]
22. Fu, Q.; Mittnik, A.; Johnson, P.L.F.; Bos, K.; Lari, M.; Bollongino, R.; Sun, C.; Giemsch, L.; Schmitz, R.; Burger, J.; et al. A revised timescale for human evolution based on ancient mitochondrial genomes. *Curr. Biol.* **2013**, *23*, 553–559. [CrossRef] [PubMed]
23. Skoglund, P.; Storå, J.; Götherström, A.; Jakobsson, M. Accurate sex identification of ancient human remains using DNA shotgun sequencing. *J. Archaeol. Sci.* **2013**, *40*, 4477–4482. [CrossRef]
24. Ralf, A.; Gonzalez, D.M.; Zhong, K.; Kayser, M. Yleaf: Software for human Y-chromosomal haplogroup inference from next-generation sequencing data. *Mol. Biol. Evol.* **2018**, *35*, 1820. [CrossRef]
25. Jun, G.; Wing, M.K.; Abecasis, G.R.; Kang, H.M. An efficient and scalable analysis framework for variant extraction and refinement from population-scale DNA sequence data. *Genome Res.* **2015**, *25*, 918–925. [CrossRef]
26. Li, H.; Handsaker, B.; Wysoker, A.; Fennell, T.; Ruan, J.; Homer, N. The sequence alignment/map format and SAMtools. *Bioinformatics* **2009**, *25*, 2078–2079. [CrossRef] [PubMed]
27. Fu, Q.; Hajdinjak, M.; Moldovan, O.T.; Constantin, S.; Mallick, S.; Skoglund, P.; Patterson, N.; Rohland, N.; Lazaridis, I.; Nickel, B.; et al. An early modern human from Romania with a recent Neanderthal ancestor. *Nature* **2015**, *524*, 216–219. [CrossRef]
28. Haak, W.; Lazaridis, I.; Patterson, N.; Rohland, N.; Mallick, S.; Llamas, B.; Brandt, G.; Nordenfelt, S.; Harney, E.; Stewardson, K.; et al. Massive migration from the steppe was a source for Indo-European languages in Europe. *Nature* **2015**, *522*, 207–211. [CrossRef]
29. Lazaridis, I.; Nadel, D.; Rollefson, G.; Merrett, D.C.; Rohland, N.; Mallick, S.; Fernandes, D.; Novak, M.; Gamarra, B.; Sirak, K.; et al. Genomic insights into the origin of farming in the ancient Near East. *Nature* **2016**, *536*, 419–424. [CrossRef]
30. Purcell, S.; Neale, B.; Todd-Brown, K.; Thomas, L.; Ferreira, M.A.R.; Bender, D.; Maller, J.; Sklar, P.; de Bakker, P.I.W.; Daly, M.J.; et al. PLINK: A tool set for whole-genome association and population-based linkage analyses. *Am. J. Hum. Genet.* **2007**, *81*, 559–575. [CrossRef]
31. Kuhn, J.M.M.; Jakobsson, M.; Günther, T. Estimating genetic kin relationships in prehistoric populations. *PLoS ONE* **2018**, *13*, e0195491.
32. Hanghøj, K.; Moltke, I.; Andersen, P.A.; Manica, A.; Korneliussen, T.S. Fast and Accurate Relatedness Estimation from High-throughput Sequencing Data in the Presence of Inbreeding. *Gigascience* **2019**, *8*, 1–9. [CrossRef]
33. Korneliussen, T.S.; Albrechtsen, A.; Nielsen, R. ANGSD: Analysis of next generation sequencing data. *BMC Bioinform.* **2014**, *15*, 356. [CrossRef]
34. McKenna, A.; Hanna, M.; Banks, E.; Sivachenko, A.; Cibulskis, K.; Kernytsky, A. The genome analysis toolkit: A MapReduce framework for analyzing next-generation DNA sequencing data. *Genome Res.* **2010**, *20*, 1297–1303. [CrossRef]
35. Genomes Project, C.; Auton, A.; Brooks, L.D.; Durbin, R.M.; Garrison, E.P.; Kang, H.M.; Korbel, J.O.; Marchini, J.L.; McCarthy, S.; McVean, G.A.; et al. A global reference for human genetic variation. *Nature* **2015**, *526*, 68–74.
36. Jacquard, A. *The Genetic Structure of Populations*; Springer-Verlag: Berlin/Heidelberg, Germany, 1974.
37. Manichaikul, A.; Mychaleckyj, J.C.; Rich, S.S.; Daly, K.; Sale, M.; Chen, W.-M. Robust relationship inference in genome-wide association studies. *Bioinformatics* **2010**, *26*, 2867–2873. [CrossRef] [PubMed]
38. Waples, R.K.; Albrechtsen, A.; Moltke, I. Allele frequency-free inference of close familial relationships from genotypes or low-depth sequencing data. *Mol. Ecol.* **2019**, *28*, 35–48. [CrossRef]
39. Chaitanya, L.; Breslin, K.; Zuñiga, S.; Wirken, L.; Pośpiech, E.; Kukla-Bartoszek, M.; Sijen, T.; de Knijff, P.; Liu, F.; Branicki, W.; et al. The HIrisPlex-S system for eye, hair and skin colour prediction from DNA: Introduction and forensic developmental validation. *Forensic Sci. Int. Genet.* **2018**, *35*, 123–135. [CrossRef]
40. Link, V.; Kousathanas, A.; Veeramah, K.; Sell, C.; Scheu, A.; Wegmann, D. ATLAS: Analysis tools for low-depth and ancient samples. *BioRxiv* **2017**, 105346. [CrossRef]
41. Davydov, E.V.; Goode, D.L.; Sirota, M.; Cooper, G.M.; Sidow, A.; Batzoglou, S. Identifying a high fraction of the human genome to be under selective constraint using GERP++. *PLoS Comput. Biol.* **2010**, *6*, e1001025. [CrossRef]
42. Walsh, S.; Chaitanya, L.; Breslin, K.; Muralidharan, C.; Bronikowska, A.; Pospiech, E.; Koller, J.; Kovatsi, L.; Wollstein, A.; Branicki, W.; et al. Global skin colour prediction from DNA. *Hum. Genet.* **2017**, *136*, 847–863. [CrossRef]
43. Walsh, S.; Chaitanya, L.; Clarisse, L.; Wirken, L.; Draus-Barini, J.; Kovatsi, L.; Maeda, H.; Ishikawa, T.; Sijen, T.; de Knijff, P.; et al. Developmental validation of the HIrisPlex system: DNA-based eye and hair colour prediction for forensic and anthropological usage. *Forensic Sci. Int. Genet.* **2014**, *9*, 150–161. [CrossRef] [PubMed]
44. Walsh, S.; Wollstein, A.; Liu, F.; Chakravarthy, U.; Rahu, M.; Seland, J.H.; Soubrane, G.; Tomazzoli, L.; Topouzis, F.; Vingerling, J.R.; et al. DNA-based eye colour prediction across Europe with the IrisPlex system. *Forensic Sci. Int. Genet.* **2012**, *6*, 330–340. [CrossRef]
45. Cariaso, M.; Lennon, G. SNPedia: A wiki supporting personal genome annotation, interpretation and analysis. *Nucleic Acids Res.* **2012**, *40*, D1308–D1312. [CrossRef] [PubMed]

46. Sawyer, S.; Krause, J.; Guschanski, K.; Savolainen, V.; Pääbo, S. Temporal patterns of nucleotide misincorporations and DNA fragmentation in ancient DNA. *PLoS ONE* **2012**, *7*, e34131. [CrossRef]
47. Paabo, S. Ancient DNA: Extraction, characterization, molecular cloning, and enzymatic amplification. *Proc. Natl. Acad. Sci. USA* **1989**, *86*, 1939–1943. [CrossRef]
48. Sikora, M.; Seguin-Orlando, A.; Sousa, V.C.; Albrechtsen, A.; Korneliussen, T.; Ko, A.; Rasmussen, S.; Dupanloup, I.; Nigst, P.R.; Bosch, M.D.; et al. Ancient genomes show social and reproductive behavior of early Upper Paleolithic foragers. *Science* **2017**, *358*, 659–662. [CrossRef]
49. Dou, J.; Sun, B.; Sim, X.; Hughes, J.D.; Reilly, D.F.; Tai, E.S.; Liu, J.; Wang, C. Estimation of kinship coefficient in structured and admixed populations using sparse sequencing data. *PLoS Genet.* **2017**, *13*, e1007021. [CrossRef] [PubMed]
50. Sinnwell, J.P.; Therneau, T.M.; Schaid, D.J. The kinship2 R package for pedigree data. *Hum. Hered.* **2014**, *78*, 91–93. [CrossRef]
51. Avent, N.D.; Reid, M.E. The Rh blood group system: A review. *Blood* **2000**, *95*, 375–387. [CrossRef]
52. Itan, Y.; Jones, B.L.; Ingram, C.J.; Swallow, D.M.; Thomas, M.G. A worldwide correlation of lactase persistence phenotype and genotypes. *BMC Evol. Biol.* **2010**, *10*, 36. [CrossRef]
53. Bersaglieri, T.; Sabeti, P.C.; Patterson, N.; Vanderploeg, T.; Schaffner, S.F.; Drake, J.A.; Rhodes, M.; Reich, D.E.; Hirschhorn, J.N. Genetic signatures of strong recent positive selection at the lactase gene. *Am. J. Hum. Genet.* **2004**, *74*, 1111–1120. [CrossRef]
54. Gerbault, P.; Liebert, A.; Itan, Y.; Powell, A.; Currat, M.; Burger, J.; Swallow, D.M.; Thomas, M.G. Evolution of lactase persistence: An example of human niche construction. *Philos. Trans. R. Soc. Lond. Ser. B Biol. Sci.* **2011**, *366*, 863–877. [CrossRef]
55. Evershed, R.P.; Payne, S.; Sherratt, A.G.; Copley, M.S.; Coolidge, J.; Urem-Kotsu, D.; Kotsakis, K.; Özdoğan, M.; Özdoğan, A.E.; Nieuwenhuyse, O.; et al. Earliest date for milk use in the Near East and southeastern Europe linked to cattle herding. *Nature* **2008**, *455*, 528–531. [CrossRef]
56. Pinhasi, R.; Fort, J.; Ammerman, A.J. Tracing the origin and spread of agriculture in Europe. *PLoS Biol.* **2005**, *3*, e410. [CrossRef] [PubMed]
57. Itan, Y.; Powell, A.; Beaumont, M.A.; Burger, J.; Thomas, M.G. The Origins of lactase persistence in Europe. *PLoS Comput. Biol.* **2009**, *5*, e1000491. [CrossRef]
58. Lomer, M.C.; Parkes, G.C.; Sanderson, J.D. Review article: Lactose intolerance in clinical practice—Myths and realities. *Aliment. Pharm.* **2008**, *27*, 93–103. [CrossRef] [PubMed]
59. Schroeder, H.; Margaryan, A.; Szmyt, M.; Theulot, B.; Wlodarczak, P.; Rasmussen, S.; Gopalakrishnan, S.; Szczepanek, A.; Konopka, T.; Jensen, T.Z.T.; et al. Unraveling ancestry, kinship, and violence in a Late Neolithic mass grave. *Proc. Natl. Acad. Sci. USA* **2019**, *116*, 10705–10710. [CrossRef]
60. Sałacińska, B.; Sałaciński, S. Rewolucja neolityczna na Mazowszu: Początki nowoczesnej gospodarki, Mazowsze. *Studia Reg.* **2010**, *5*, 51–72. (In Polish)
61. Szmyt, M. Kugelamphoren-Gemeinschaften in Mittel- und Osteuropa: Siedlungs-Strukturen und Soziale Fragen. *UPA* **2002**, *90*, 195–233. (In German)
62. Szmyt, M. Wędrówki bliskie i dalekie. Ze studiów nad organizacją społeczną i gospodarką ludności kultury amfor kulistych na terenie Europy Środkowej i Wschodniej. In *Nomadyzm i Pastoralizm w Międzyrzeczu Wisły i Dniepru (Neolit, Eneolit, Epoka Brązu), Archeologia Bimaris. Dyskusje 3*; Kośko, A., Szmyt, M., Eds.; Wydawnictwo Poznańskie: Poznań, Poland, 2004; pp. 117–136. (In Polish)
63. Woidich, M. *Die Westliche Kugelamphorenkultur: Untersuchungen zu Ihrer Raumzeitlichen Differenzierung, Kulturellen und Anthropologischen Identitét*; De Gruyter: Berlin, Germany, 2014; 442p. (In German)

Article

A Revised Model of Anatomically Modern Human Expansions Out of Africa through a Machine Learning Approximate Bayesian Computation Approach

Maria Teresa Vizzari, Andrea Benazzo, Guido Barbujani and Silvia Ghirotto *

Department of Life Sciences and Biotechnology, University of Ferrara, 44121 Ferrara, Italy; mariateresa.vizzari@unife.it (M.T.V.); andrea.benazzo@unife.it (A.B.); guido.barbujani@unife.it (G.B.)
* Correspondence: silvia.ghirotto@unife.it

Received: 5 November 2020; Accepted: 14 December 2020; Published: 16 December 2020

Abstract: There is a wide consensus in considering Africa as the birthplace of anatomically modern humans (AMH), but the dispersal pattern and the main routes followed by our ancestors to colonize the world are still matters of debate. It is still an open question whether AMH left Africa through a single process, dispersing almost simultaneously over Asia and Europe, or in two main waves, first through the Arab Peninsula into southern Asia and Australo-Melanesia, and later through a northern route crossing the Levant. The development of new methodologies for inferring population history and the availability of worldwide high-coverage whole-genome sequences did not resolve this debate. In this work, we test the two main out-of-Africa hypotheses through an Approximate Bayesian Computation approach, based on the Random-Forest algorithm. We evaluated the ability of the method to discriminate between the alternative models of AMH out-of-Africa, using simulated data. Once assessed that the models are distinguishable, we compared simulated data with real genomic variation, from modern and archaic populations. This analysis showed that a model of multiple dispersals is four-fold as likely as the alternative single-dispersal model. According to our estimates, the two dispersal processes may be placed, respectively, around 74,000 and around 46,000 years ago.

Keywords: approximate Bayesian computation; demographic history; human evolution; migration; machine learning; random forest; whole-genome data

---

1. Introduction

Levels and patterns of genome diversity reflect past demographic processes, and a crucial turning point in our demographic history is the expansion of anatomically modern humans (AMH) from Africa. Some aspects of this process seem rather well established. First, what is often called the ancestral African population should not be regarded as a single, biologically homogeneous unit, but as a structured population hosting regional diversity [1]. Second, the AMH expansion was accompanied by the disappearance of preexisting archaic human forms [2,3] Third, a variable component of the genomes of most present populations—always small, seldom zero—comes from anatomically archaic ancestors [4].

Conversely, there is disagreement over other aspects of the AMH expansion out of Africa, such as the number of major dispersal events, their timing, and the geographical routes followed by migrating people. Groups of AMH may have left Africa more than 100,000 years ago [5], but genetic evidence suggests that such early phenomena were not successful and did not lead to the establishment of permanent non-African populations. One expansion left traces in modern genomes; it took place between 60,000 and 50,000 years ago, along a Northern route in the Nile valley and across the

Near East (see e.g., [6–8]). However, based on cranial morphology, Lahr and Foley [9] proposed an additional, earlier migration through a Southern route, from the Horn of Africa into the Arab peninsula, Southern Asia, and Australo-Melanesia. We shall refer to these alternative models as Single Dispersal (SD) and Multiple Dispersal (MD) hypotheses. The MD hypothesis found support in several studies, and notably in a comparison of cranial and DNA diversity data [10] but broader genomic analyses gave contradictory results. Tassi and colleagues [11] and, to a lesser extent, Pagani et al. [12] described patterns consistent with two dispersal processes, the first one overlapping in time with the proposed early Southern exit from Africa [11]. On the other hand, two studies of different genomic datasets concluded that there is little [4] or no evidence [13] for such an early dispersal process, and hence that AMH either left Africa in a single major migrational wave, or perhaps in several waves, but then only one of them contributed to the ancestry of modern populations.

Malaspinas et al. [13] conclusion in favor of SD was not really based on an explicit comparison between models. In their paper, indeed, they considered an MD model in which East Asians and Europeans have a more recent common ancestor than Aboriginal Australians and East Asians. and they estimated the models' parameters. The evidence supporting the SD model came from the overlapping estimation for the divergence times of the ancestors of Aboriginal Australians and Eurasians.

This non-straightforward procedure was due to an implicit limitation of the composite likelihood method they applied, in which model selection may be performed through likelihood ratio tests (LRT) or by the Akaike Information Criterion (AIC; [14,15]). LRT and AIC can only be used to understand which modifications significantly improve the model, without explicit model testing and a direct attribution of probabilities to each tested scenario.

To understand which model, SD or MD, better accounts for the current levels of genome diversity, in this study we formally compare them by a recently developed Approximate Bayesian Computation framework, based on the study of the observed Frequency Distributions of four categories of Segregating Sites for pair of populations (FDSS) [16]. ABC is a powerful and flexible framework, based on computer simulations, to perform model selection and estimate models' parameters. In its original formulation [17,18] the ABC algorithm suffered from two main issues, related to the simulation effort and to the number of summary statistics used to summarize the data. These issues limited the possibility to use ABC for the analysis of complex demographic histories and/or large datasets. In 2015, the introduction of a paradigm shift in the ABC model selection procedure based on a Machine Learning approach called Random Forest (ABC-RF, [19]), allowed to overcome the above-cited limitations and paved the ground for the application of ABC to the study of complex models through the analysis of complete genomes. Under ABC-RF, the model selection procedure is rephrased as a classification problem. At first, the classifier is constructed from simulations from the prior distribution via a machine learning RF algorithm. Once the classifier is constructed and applied to the observed data, the posterior probability of the resulting model can be approximated through another RF that regresses the selection error over the statistics used to summarize the data. The number of simulations necessary to obtain reliable estimates passed from a few million to a few thousand; the informative statistics are systematically extracted from the pool used to summarize the data. In 2018, a similar approach, based on a machine-learning tool of regression RF, has been developed for parameter estimation [20]. In [16] we showed that the ABC-RF algorithm, combined with the inferential power provided by the FDSS, can be satisfactorily exploited to estimated past population dynamics even in case of complex demographic histories, thus making the approach particularly suitable to the analysis of SD and MD models.

Under both SD and MD models, the structure of the past populations is the same, but the tree topologies differ in that they assume, respectively, one ancestral population for the SD model, and two ancestral populations leaving Africa at different times for the MD model. As the Australo-Melanesian represent the population that might carry the signal of the first wave of migrations out of the African continent and also, to make sure that the different results obtained by [12,13] were not due to differences in the Australo-Melanesian samples available, we repeated our analyses considering genomes coming from both studies, obtaining results that seem consistent and informative.

2. Materials and Methods

2.1. The FDSS

We summarized the data through the FDSS, i.e., the frequency distributions of the four mutually exclusive categories of segregating sites for pair of populations (i.e., private polymorphisms in either population, shared polymorphisms, and fixed differences [21]). This statistic proved to be powerful for reconstructing even a complex series of demographic processes [16]. The FDSS is calculated considering each genome analyzed as subdivided into a certain number of independent fragments of a certain length, and for each fragment, the number of sites belonging to each of the four above-mentioned categories is counted. The final vector of summary statistics is thus composed by the truncated frequency distribution of fragments having from 0 to n segregating sites in each category, for each pair of populations considered. We fixed the maximum number of segregating sites in a locus of a certain length to 100, and hence the last category contains all the observations higher than 100.

We calculated the FDSS using a python script (available on Github https://github.com/anbena/ABC-FDSS) [16]. The ABC-RF model selection estimates have been obtained using the function *abcrf* from the package *abcrf* and employing a forest of 500 classification trees, a number suggested providing the best trade-off between computational efficiency and statistical precision [19]. Before proceeding with the model selection procedure, we computed the confusion matrices and evaluated the out-of-bag classification error (CE) and the proportion of True Positives (1-CE), which are representative of the power of the whole inferential procedure. The ABC-RF parameters estimation on the most supported models have been performed through the function *regAbcrf* from the package *abcrf* and employing a forest of 500 regression trees. An outline of our entire workflow is reported in Figure S1.

2.2. Simulated Models of Anatomically Modern Humans Expansion Out of Africa

We tested two alternative models of expansion of anatomically modern humans out of the African continent (Figure 1), both sharing the same structure for the archaic groups, but differing for the relationships among modern populations. To design the models, we followed the parametrization proposed by [13], with some modifications detailed below. The first model (SD) indeed accounts for a single dispersal from Africa giving rise to both modern Eurasians and Australo-Melanesians, the second model (MD) accounts for two different waves of migrations, from two different African source populations, giving rise, first, to the modern Australo-Melanesians and, later to the modern Eurasians. The archaic groups consist of three Denisovan populations, two Neanderthal populations, and an unknown archaic population ancestral to both Neandertals and Denisovans. We explicitly considered admixture pulses from archaic to modern populations: a pulse from the archaic unknown population to Australo-Melanesians (as reported in [22]), two pulses from two different Denisovan populations to Asians and Australo-Melanesians [23,24], two pulses from the same Neandertal population to modern humans just after the separation between African and non-African populations, and to the ancestor of all Eurasians [25–27]. Both models account for the presence of a Basal European population, as described in [28–30]. This (so far, unknown) population contributed genes to modern Europeans, possibly diluting the contribution of archaic Neandertal variants in European genomes. The SD and MD models have 45 and 50 free parameters (i.e., parameters whose values are defined by prior distributions), respectively. The prior distributions associated with these parameters were set following what was proposed in the recent literature by [13,23,30], and are reported in Tables S1 and S2. We considered a generation time of 29 years, and we fixed the mutation rate at 1.25×10^{-8} bp/generation [31] and the intra-locus recombination rate at 1.12×10^{-8}, all values as in [13].

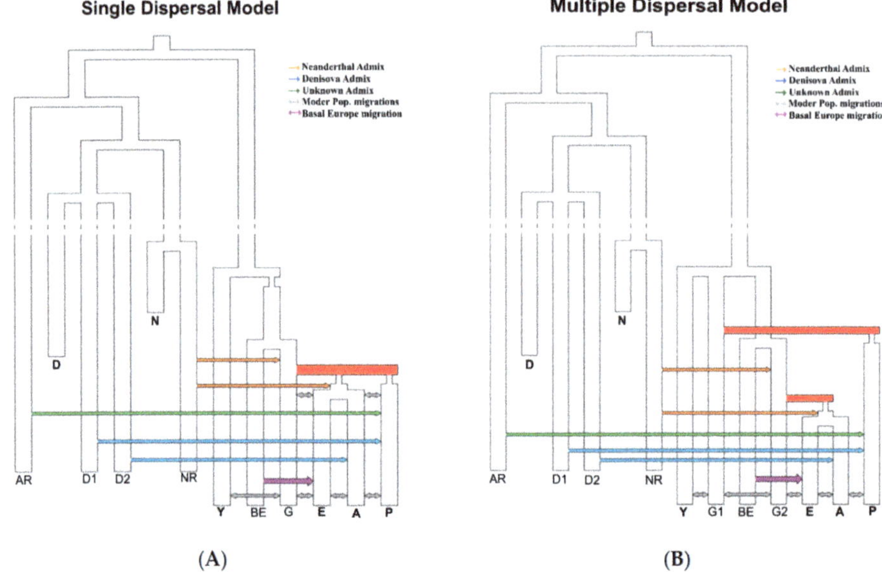

Figure 1. Demographic models compared: Single Dispersal (**A**) and Multiple Dispersals (**B**). AR: unknown archaic population; D-D1-D2: Denisovan groups; N-NR: Neandertal and Neandertal related groups; Y: African population; G1-G2: ghost populations; BE: Basal Europe population; E: European population; A: Asian population; P: Australo-Melanesian population.

We performed 20,000, 50,000, and 100,000 simulations for each model with *ms* [32], to evaluate the Prior Error Rate and identify the optimum number of simulations to use. At each iteration, we sampled six diploid genomes, one Neandertal, one Denisova, one African, one European, one Asian, and one Papuan. The FDSS was calculated from 10,000 independent genomic fragments of 500 bp length.

2.3. Observed Genomic Data

We analyzed the high-coverage genomes of Denisova [33] and Neandertal [26], together with worldwide modern human samples from [12]. All the individuals were mapped against the human reference genome *hg19* build 37. To calculate the observed *FDSS* we only considered autosomal regions outside known and predicted genes ± 10,000 bp and outside CpG islands and repeated regions (as defined on the UCSC platform, [34]). We extracted 10,000 independent fragments of 500 bp length, separated by at least 10,000 bps in genomic regions that passed a set of minimal quality filters used for the analysis of the ancient genomes (*map35_50%*; [26,33]). We also included in the analysis of the 25 Papuan individuals published by [13]. For these individuals, we downloaded the alignments in CRAM format from https://www.ebi.ac.uk/ega/datasets/EGAD00001001634. The *mpileup* and *call* commands from *samtools-1.6* [35], were used to call all variants within the 10,000 neutral genomic fragments, using the –consensus-caller flag, without considering indels. We then filtered the initial call set according to the filters reported in [13] using *vcflib* and *bcftools* [35]. The complete set of samples used for the comparison between SD and MD are reported in Table S3.

In each models' comparison, we evaluated the genomic variation of one Denisova, one Neandertal, one African (Congo-pygmies), one European (Estonians), one Asian (Vietnamese), and one Australo-Melanesian (Papuans). We decided to restrict the analysis to one high coverage diploid genome per population since previous extensive analyses showed that a single individual sampled per population has a comparable discrimination power as twenty chromosomes [16]. However, to ensure the consistency of the results, we performed several model selection procedures (a) taking into account at each run one out of six Papuans from [12] or one of 25 Papuans from [13]; (b) considering alternative individuals as representative of African, European, and Asian populations (Table S4).

2.4. Assessment of the Quality of the Parameters Estimated

One of the most interesting features of ABC is its high flexibility for model checking, i.e., for assessing the quality of the estimates inferred from real data. This is mainly achieved through the analysis of pseudo-observed data (pods), i.e., simulated datasets generated under known conditions. To determine whether the observed data would contain enough information to estimate parameters of the multi-dimensional model tested, we exploited 1000 pods, each generated from the most supported model (i.e., the MD model) and through a known combination of demographic parameters. Using these pods, for each parameter we calculated the following indices:

- The coefficient of determination (R^2). R^2 is the fraction of variance of the parameters explained by the summary statistics used to build the regression model. In the absence of an established threshold value, there is a general agreement that when $R^2 < 0.10$, the summary statistics do not convey enough information about the parameter estimates [36].
- The relative bias. To calculate the relative bias, we estimated the parameters for each pod with the same approach used for the observed data. The bias depends on the sum of differences between the 1000 estimates of each parameter thus obtained and the known (true) value, and it is calculated as

$$\frac{1}{n}\sum_{i=1}^{n}\frac{\theta_i - \theta}{\theta}$$

where θ_i is the estimator of the parameter θ (true value), and n is the number of pods used (1000 in our case). Because bias is relative, a value of 1 corresponds to a bias equal to 100% of the true value.
- The root mean square error (RMSE). To calculate the RMSE we re-estimated parameters using pods. The RMSE depends the sum of squared differences between the 1000 estimates of each parameter thus obtained and the true value and it is calculated as:

$$\sqrt{\frac{1}{n}\sum_{i=1}^{n}(\theta_i - \theta)^2}$$

- The factor 2, representing the proportion of the 1000 estimated median values lying between 50% and 200% of the true value.
- The 50% and 90% coverage, defined as the proportion of times that the known value lies within the 50% and the 90% credible interval of the 1000 estimates.

3. Results

3.1. Model Selection

Table 1 and Table S5 show the results of the power check of the comparison between SD and MD. Predictably, the Prior Error rate, which indicates the global quality of the ML classifier, decreases for increasing numbers of simulations in the reference table (from 20,000 to 100,000); for this reason, we decided to use 100,000 simulations for the subsequent analyses. The proportion of True Positives, that is the proportion of times the SD or the MD model is correctly recognized by the model selection procedure, is above 70% for both SD and MD, with a mean posterior probability associated with the true demography of about 75%.

Table 1. Power test for model comparison using a reference table with 100,000 simulations per model.

Prior Err. Rate	True Positive SD	True Positive MD	Post. Prob. SD	Post. Prob. MD
0.26	0.73	0.75	0.75	0.73

Table 2 and Table S4 show the results of the model selection. Regardless of the Papuan individual considered, and the combination of non-Australo-Melanesian tested, the model selection analyses supported the MD model as the scenario best explaining the recent evolution of anatomically modern humans out of Africa, with probabilities ranging from 78 to 84%.

Table 2. Model Selection results using Papuan individuals from [12,13]. In the first column are reported the ID of the Papuan samples used for the model choice. The second column shows the model selected by the ABC procedure. In the third and the fourth columns are reported the votes assigned to the SD and MD models by the Random-Forest algorithm. The last column shows the posterior probabilities associated with the most supported model. The samples with the highest posterior probabilities (in bold) were selected to perform the parameter estimation of the MD model.

ID_Individual	Selected Model	Votes SD	Votes MD	Post. Prob.
EGAN00001279031	MD	94	406	0.822
EGAN00001279039	MD	86	414	0.806
EGAN00001279047	MD	111	389	0.798
EGAN00001279054	MD	128	372	0.809
EGAN00001279032	**MD**	**90**	**410**	**0.825**
EGAN00001279040	MD	113	387	0.784
EGAN00001279048	MD	99	401	0.805
EGAN00001279033	MD	108	392	0.791
EGAN00001279041	MD	111	389	0.797
EGAN00001279049	MD	126	374	0.789
EGAN00001279034	MD	150	350	0.797
EGAN00001279042	MD	109	391	0.791
EGAN00001279050	MD	111	389	0.797
EGAN00001279035	MD	108	392	0.799
EGAN00001279043	MD	97	403	0.802
EGAN00001279051	MD	117	383	0.786
EGAN00001279036	MD	136	364	0.778
EGAN00001279044	MD	109	391	0.784
EGAN00001279052	MD	100	400	0.815
EGAN00001279037	MD	96	404	0.800
EGAN00001279045	MD	148	352	0.787
EGAN00001279053	MD	100	400	0.796
EGAN00001279038	MD	91	409	0.811
EGAN00001279046	MD	104	396	0.781
EGAN00001279055	MD	138	362	0.787
Koinb1	MD	165	335	0.810
Koinb2	MD	129	371	0.811
Koinb3	MD	175	325	0.820
Kosip1	MD	152	348	0.818
Kosip2	MD	136	364	0.788
Kosip3	**MD**	**123**	**377**	**0.830**

3.2. Parameters Estimation

Once identified the MD as the most probable model, we moved to estimate its parameter values maximizing the fit between observed and simulated genomic data. To do this, we exploited the recently developed ML method, based on a regression RF approach [20]. As detailed in [20], a faithful estimation of parameters' posterior distribution may be now achieved with a reduced number of simulations (i.e., a few thousand; we used 100,000 simulations), making it feasible to also perform an accurate assessment of the quality of the parameters estimated using pods.

Parameters were estimated from two observed datasets (one with a Papuan individual from [13] and one with a Papuan individual from [12]), those which produced the highest value of posterior probability for the MD model in the model selection (Tables 3 and 4). The posterior plots and the definition of the parameter's acronyms are reported in Supplementary Materials (Figures S2–S10, Table S6). The R^2, the bias, the RMSE, the Factor 2, and the 50–90% Coverage associated with each of these parameters are shown in Table 5. As expected for complex demography, many parameters are not well estimated, as indicated by low R^2, high bias, and high RMSE. The parameters showing better estimation quality are the effective population sizes, in particular those associated with the ancestral population of African and non-African modern humans (nYG, R^2 = 91%), and the ancestral population of modern and archaic groups (nAM, R^2 = 99%). The divergence times appear to have been estimated reasonably well, with most of R^2s above 10%. This is true in particular for the times of the two Out of Africa events, which also show a low bias and a high Factor2 and Coverage. On the other hand, it is evident that the data tell us very little about admixture events (their timing and admixture proportions) and migration rates. Although disappointing, this is not unexpected, and high levels of uncertainty associated with these parameters were already reported [13].

Table 3. Estimated parameters for the MD model using the Papuan samples from [13]. The mean and the median estimated values are listed, as well as the 90% and the 50% credible intervals. The parameters cited in the text are reported in bold.

Parameter	Mean	Median	Variance	Q (0.05)	Q (0.95)	Q (0.25)	Q (0.75)
nAR	**2822**	**2793**	**5.77×10^4**	**2540**	**3410**	**2666**	**2914**
nY	**19,077**	**14,347**	**1.72×10^8**	**4204**	**44,993**	**7976**	**29117**
nG1	26,191	26,995	**2.08×10^8**	3253	47,385	13,670	39,819
nG2	23,473	22,275	**1.96×10^8**	1903	46,649	11,151	34,663
nBE	25,612	26,269	**2.08×10^8**	2731	47,604	13,394	38,160
nE	**13,498**	**6616**	**2.07×10^8**	**627**	**42,565**	**1616**	**23,761**
nA	**16,360**	**11,553**	**2.25×10^8**	**773**	**44,620**	**2599**	**28,065**
nP	**24,268**	**24,839**	**2.34×10^8**	**1535**	**47,534**	**10,756**	**37,349**
nYG	**23,317**	**22,292**	**3.19×10^7**	**17,112**	**35,456**	**19,789**	**25,425**
nNNR	**2424**	**2343**	**1.22×10^5**	**2057**	**3001**	**2219**	**2504**
nDDR	21,360	19,680	2.00×10^8	1570	46,512	9482	32,332
nDN	17,025	12,576	1.77×10^8	2789	43,117	5312	27,001
nADN	19,733	16,531	2.28×10^8	2108	47,465	5770	31,455
nAM	**18,846**	**18,745**	**1.73×10^6**	**16,780**	**21,023**	**17,911**	**19,745**
rP	0.0214	0.0146	8.36×10^{-4}	0.0105	0.0532	0.0119	0.0192
rEA	0.0313	0.0179	1.91×10^{-3}	0.0109	0.0869	0.0142	0.0303
tdYG1	**101,162**	**103,842**	**7.61×10^8**	**54,830**	**140,536**	**78,262**	**125,226**
tdYG2	**99,000**	**98,925**	**7.13×10^8**	**55,038**	**137,970**	**76,482**	**124,250**
tdOA1	**77,106**	**73,566**	**5.86×10^8**	**47,019**	**120,206**	**55,392**	**96,881**
tOAbot1	73,389	66,248	6.14×10^8	44,341	118,942	52,082	93,165
tdOA2	**47,524**	**45,937**	**3.99×10^7**	**40,394**	**59,245**	**42,597**	**51,019**
tOAbot2	45,223	43,282	5.30×10^7	37,718	58,387	40,110	48,153
tdG2BE	68,415	61,497	3.78×10^8	50,281	113,560	53,713	75,889
tdEA	**38,187**	**37,017**	**4.33×10^7**	**30,483**	**50,076**	**33,374**	**41,444**
taNG2	52,032	49,731	8.13×10^7	42,680	69,758	45,402	55,444
taNEA	41,663	40,005	4.51×10^7	33,965	55,743	36,653	45,055
taARP	61,567	55,048	4.53×10^8	37,831	106,642	43,945	75,654
taD1P	51,047	44,460	3.89×10^8	31,094	95,155	36,207	58,088
taD2A	28,645	27,059	4.24×10^7	20,958	39,746	23,730	32,456
taBEE	25,269	24,844	1.00×10^8	11,194	45,254	16,827	31,380
paNG2	5.19×10^{-2}	4.99×10^{-2}	7.71×10^{-4}	9.44×10^{-3}	9.52×10^{-3}	2.91×10^{-2}	7.73×10^{-2}
paNEA	4.73×10^{-2}	4.73×10^{-2}	7.95×10^{-4}	5.36×10^{-3}	9.57×10^{-2}	2.30×10^{-2}	7.01×10^{-2}

Table 3. *Cont.*

Parameter	Mean	Median	Variance	Q (0.05)	Q (0.95)	Q (0.25)	Q (0.75)
paARP	4.82×10^{-2}	4.83×10^{-2}	9.00×10^{-4}	4.97×10^{-3}	9.45×10^{-2}	2.09×10^{-2}	7.71×10^{-2}
paD1P	5.21×10^{-2}	5.27×10^{-2}	8.43×10^{-4}	4.58×10^{-3}	9.53×10^{-2}	2.84×10^{-2}	7.85×10^{-2}
paD2A	4.74×10^{-2}	4.72×10^{-2}	8.46×10^{-4}	3.95×10^{-3}	9.32×10^{-2}	2.17×10^{-2}	7.24×10^{-2}
paBEE	2.78×10^{-1}	2.85×10^{-1}	1.61×10^{-2}	6.83×10^{-2}	4.79×10^{-1}	1.71×10^{-1}	3.83×10^{-1}
mYG1	4.75×10^{-4}	4.62×10^{-4}	9.64×10^{-8}	2.61×10^{-5}	9.48×10^{-4}	1.92×10^{-4}	7.54×10^{-4}
mG1Y	4.74×10^{-4}	4.64×10^{-4}	7.95×10^{-8}	4.65×10^{-5}	9.30×10^{-4}	2.25×10^{-4}	6.98×10^{-4}
mG1G2	4.93×10^{-4}	4.80×10^{-4}	8.50×10^{-8}	4.54×10^{-5}	9.41×10^{-4}	2.49×10^{-4}	7.63×10^{-4}
mG2G1	5.34×10^{-4}	5.61×10^{-4}	8.83×10^{-8}	4.77×10^{-5}	9.68×10^{-4}	2.69×10^{-4}	7.94×10^{-4}
mG2E	5.23×10^{-4}	5.29×10^{-4}	8.13×10^{-8}	5.19×10^{-5}	9.57×10^{-4}	2.84×10^{-4}	7.81×10^{-4}
mEG2	4.21×10^{-4}	3.69×10^{-4}	7.78×10^{-8}	3.73×10^{-5}	9.07×10^{-4}	1.85×10^{-4}	6.48×10^{-4}
mEA	4.19×10^{-4}	3.60×10^{-4}	8.63×10^{-8}	3.73×10^{-5}	9.66×10^{-4}	1.81×10^{-4}	6.45×10^{-4}
mAE	5.33×10^{-4}	5.69×10^{-4}	7.63×10^{-8}	5.82×10^{-5}	9.33×10^{-4}	2.90×10^{-4}	7.57×10^{-4}
mAP	1.70×10^{-4}	1.27×10^{-4}	2.26×10^{-8}	1.42×10^{-5}	5.16×10^{-4}	7.40×10^{-5}	2.10×10^{-4}
mPA	1.28×10^{-4}	1.02×10^{-4}	1.18×10^{-8}	8.01×10^{-6}	3.37×10^{-4}	4.52×10^{-5}	1.72×10^{-4}
m1G2EA	4.96×10^{-4}	5.01×10^{-4}	8.24×10^{-8}	5.60×10^{-6}	9.47×10^{-4}	2.45×10^{-4}	7.53×10^{-4}
m1EAG2	4.46×10^{-4}	4.00×10^{-4}	8.23×10^{-8}	5.18×10^{-5}	9.49×10^{-4}	1.99×10^{-4}	6.95×10^{-4}
m1EAP	4.25×10^{-4}	3.97×10^{-4}	7.57×10^{-8}	2.77×10^{-5}	9.07×10^{-4}	1.95×10^{-4}	6.39×10^{-4}
m1PEA	4.40×10^{-4}	4.02×10^{-4}	8.39×10^{-8}	4.04×10^{-5}	9.31×10^{-4}	1.77×10^{-4}	6.93×10^{-4}

Table 4. Estimated parameters for the MD model using the Papuan samples from [12]. The mean and the median estimated values are listed, as well as the 90% and the 50% credible intervals. The parameters cited in the text are reported in bold.

Parameter	Mean	Median	Variance	Q (0.05)	Q (0.95)	Q (0.25)	Q (0.75)
nAR	**2803**	**2783**	**4.57×10^{4}**	**2532**	**3302**	**2668**	**2900**
nY	**19,182**	**14,771**	**1.62×10^{8}**	**4379**	**44,930**	**8223**	**29,102**
nG1	26,722	28,003	2.18×10^{8}	2702	47,514	14,075	40,579
nG2	25,325	27,394	1.97×10^{8}	2218	47,188	13,362	36,308
nBE	25,684	26,296	2.17×10^{8}	2194	47,896	13,706	38,919
nE	**12,485**	**5373**	**1.94×10^{8}**	**699**	**42,194**	**1616**	**21,836**
nA	**14,543**	**8978**	**2.10×10^{8}**	**916**	**43,930**	**2214**	**26,207**
nP	**19,089**	**16,639**	**2.16×10^{8}**	**1048**	**46,319**	**4980**	**30,429**
nYG	**22,857**	**21,922**	**2.62×10^{7}**	**17,112**	**31,789**	**19,579**	**25,130**
nNNR	**2422**	**2336**	**1.24×10^{5}**	**2057**	**3023**	**2219**	**2531**
nDDR	21,778	20,572	1.94×10^{8}	1640	46291	9606	32,332
nDN	16,239	11,846	1.59×10^{8}	2879	41321	5311	25,523
nADN	19,279	16,531	2.21×10^{8}	2108	47070	4884	31,082
nAM	**18,629**	**18,574**	**1.57×10^{6}**	**16,671**	**20,691**	**17,779**	**19,476**
rP	0.0215	0.0143	6.10×10^{-4}	0.0104	0.0576	0.0118	0.0204
rEA	0.0314	0.0179	1.94×10^{-3}	0.0109	0.0869	0.0144	0.0310
tdYG1	**98,829**	**99,987**	**7.31×10^{8}**	**54,220**	**140,009**	**76,337**	**122,428**
tdYG2	**97,430**	**96,686**	**6.87×10^{8}**	**54,693**	**138,490**	**76,482**	**120,370**
tdOA1	**74,244**	**68,987**	**5.32×10^{8}**	**46,663**	**119,539**	**54,334**	**89,685**
tOAbot1	70,341	64,285	5.47×10^{8}	43,471	116,608	50,992	85,938
tdOA2	**48,554**	**46,257**	**7.36×10^{7}**	**40,559**	**64,865**	**42,739**	**51,453**
tOAbot2	46,366	43,475	8.49×10^{7}	37,922	63,074	40,247	50,084
tdG2BE	68,122	62,035	3.36×10^{8}	50,281	105,774	53,533	76,526
tdEA	**37,747**	**35,936**	**5.05×10^{7}**	**30,381**	**50,399**	**32,690**	**40,845**
taNG2	53,606	50,116	1.08×10^{8}	43,274	73,012	46,917	57,484
taNEA	42,255	40,175	7.98×10^{7}	33,449	56,376	37,030	45,231
taARP	61,203	54,697	4.60×10^{8}	37,428	106,643	43,994	73,444
taD1P	48,493	43,651	2.90×10^{8}	31,343	86,579	36,450	55,023
taD2A	29,298	27,601	5.05×10^{7}	21,090	41,451	24,133	32,700
taBEE	23,871	23,356	9.64×10^{7}	10,508	40,711	15,268	30,666

Table 4. *Cont.*

Parameter	Mean	Median	Variance	Q (0.05)	Q (0.95)	Q (0.25)	Q (0.75)
paNG2	5.29×10^{-2}	5.35×10^{-2}	7.32×10^{-4}	8.94×10^{-3}	9.52×10^{-2}	3.18×10^{-2}	7.51×10^{-2}
paNEA	5.12×10^{-2}	5.22×10^{-2}	7.83×10^{-4}	5.58×10^{-3}	9.60×10^{-2}	2.69×10^{-2}	7.44×10^{-2}
paARP	5.02×10^{-2}	5.06×10^{-2}	8.74×10^{-4}	5.45×10^{-3}	9.49×10^{-2}	2.36×10^{-2}	7.81×10^{-2}
paD1P	5.23×10^{-2}	5.50×10^{-2}	8.00×10^{-4}	6.13×10^{-3}	9.41×10^{-2}	2.78×10^{-2}	7.66×10^{-2}
paD2A	4.82×10^{-2}	4.52×10^{-2}	8.87×10^{-4}	4.93×10^{-3}	9.58×10^{-2}	2.27×10^{-2}	7.39×10^{-2}
paBEE	2.79×10^{-1}	2.91×10^{-1}	1.65×10^{-2}	6.58×10^{-2}	4.78×10^{-1}	1.68×10^{-1}	3.88×10^{-1}
mYG1	4.47×10^{-4}	4.08×10^{-4}	8.52×10^{-8}	3.74×10^{-5}	9.32×10^{-4}	1.89×10^{-4}	6.97×10^{-4}
mG1Y	4.92×10^{-4}	4.91×10^{-4}	7.55×10^{-8}	5.11×10^{-5}	9.27×10^{-4}	2.79×10^{-4}	7.28×10^{-4}
mG1G2	4.74×10^{-4}	4.59×10^{-4}	8.40×10^{-8}	4.41×10^{-5}	9.35×10^{-4}	2.31×10^{-4}	7.32×10^{-4}
mG2G1	5.20×10^{-4}	5.23×10^{-4}	9.07×10^{-8}	4.77×10^{-5}	9.67×10^{-4}	2.34×10^{-4}	7.93×10^{-4}
mG2E	5.16×10^{-4}	5.29×10^{-4}	7.87×10^{-8}	5.67×10^{-5}	9.55×10^{-4}	2.85×10^{-4}	7.60×10^{-4}
mEG2	3.77×10^{-4}	3.04×10^{-4}	8.13×10^{-8}	2.70×10^{-5}	9.11×10^{-4}	1.30×10^{-4}	5.80×10^{-4}
mEA	5.07×10^{-4}	5.15×10^{-4}	8.78×10^{-8}	4.74×10^{-5}	9.57×10^{-4}	2.52×10^{-4}	7.68×10^{-4}
mAE	4.67×10^{-4}	4.68×10^{-4}	7.94×10^{-8}	1.78×10^{-5}	9.17×10^{-4}	2.29×10^{-4}	7.07×10^{-4}
mAP	5.17×10^{-4}	5.12×10^{-4}	7.28×10^{-8}	1.04×10^{-4}	9.35×10^{-4}	2.78×10^{-4}	7.50×10^{-4}
mPA	4.05×10^{-4}	3.79×10^{-4}	5.71×10^{-8}	5.15×10^{-5}	8.70×10^{-4}	2.27×10^{-4}	5.41×10^{-4}
m1G2EA	5.20×10^{-4}	5.21×10^{-4}	8.85×10^{-8}	4.88×10^{-5}	9.74×10^{-4}	2.74×10^{-4}	7.90×10^{-4}
m1EAG2	4.56×10^{-4}	4.30×10^{-4}	7.91×10^{-8}	5.77×10^{-5}	9.24×10^{-4}	2.09×10^{-4}	7.16×10^{-4}
m1EAP	4.92×10^{-4}	5.12×10^{-4}	7.88×10^{-8}	6.32×10^{-5}	9.42×10^{-4}	2.47×10^{-4}	7.11×10^{-4}
m1PEA	4.78×10^{-4}	4.59×10^{-4}	7.42×10^{-8}	6.17×10^{-5}	9.24×10^{-4}	2.44×10^{-4}	7.02×10^{-4}

Table 5. Accuracy of the estimated parameters of the MD model assessed by 1000 pods. The parameters cited in the text are reported in bold.

Parameters	R^2	Bias	RMSE	Factor 2	Coverage 90%	Coverage 50%
nAR	**0.84**	**−0.0020**	$\mathbf{5.90 \times 10^{3}}$	**0.990**	**0.935**	**0.553**
nY	**0.54**	**0.1900**	$\mathbf{1.04 \times 10^{4}}$	**0.867**	**0.919**	**0.522**
nG1	0.08	2.0020	1.46×10^{4}	0.702	0.880	0.466
nG2	0.17	0.9175	1.36×10^{4}	0.698	0.915	0.497
nBE	0.02	2.2194	1.47×10^{4}	0.722	0.895	0.479
nE	**0.33**	**0.4278**	$\mathbf{1.25 \times 10^{4}}$	**0.767**	**0.908**	**0.523**
nA	**0.28**	**0.4159**	$\mathbf{1.20 \times 10^{4}}$	**0.795**	**0.922**	**0.532**
nP	**0.39**	**0.3425**	$\mathbf{1.21 \times 10^{4}}$	**0.791**	**0.908**	**0.501**
nYG	**0.91**	**0.0020**	$\mathbf{3.54 \times 10^{3}}$	**0.998**	**0.957**	**0.650**
nNNR	**0.92**	**0.0086**	$\mathbf{3.64 \times 10^{3}}$	**0.998**	**0.966**	**0.622**
nDDR	0.36	0.3529	1.18×10^{4}	0.800	0.923	0.522
nDN	0.54	0.1979	1.09×10^{4}	0.842	0.941	0.534
nADN	0.33	0.7749	1.29×10^{4}	0.705	0.930	0.476
nAM	**0.99**	**0.0067**	$\mathbf{5.40 \times 10^{2}}$	**0.997**	**0.995**	**0.870**
rP	0.10	0.1110	6.79×10^{-2}	0.721	0.879	0.521
rEA	0.10	0.0983	5.65×10^{-2}	0.748	0.915	0.547
tdYG1	**0.25**	**0.0629**	$\mathbf{2.23 \times 10^{4}}$	**0.998**	**0.928**	**0.576**
tdYG2	**0.25**	**0.0630**	$\mathbf{2.25 \times 10^{4}}$	**0.996**	**0.934**	**0.573**
tdOA1	**0.19**	**0.0025**	$\mathbf{1.99 \times 10^{4}}$	**0.998**	**0.911**	**0.540**
tOAbot1	0.19	0.0052	1.99×10^{4}	0.996	0.918	0.544
tdOA2	**0.13**	**−0.0257**	$\mathbf{1.24 \times 10^{4}}$	**0.998**	**0.883**	**0.511**
tOAbot2	0.13	−0.0261	1.24×10^{4}	0.995	0.881	0.512
tdG2BE	0.16	−0.0016	1.98×10^{4}	0.999	0.913	0.523
tdEA	**0.08**	**−0.0167**	$\mathbf{9.09 \times 10^{3}}$	**0.989**	**0.898**	**0.495**
taD2A	0.04	0.0116	7.35×10^{3}	0.993	0.905	0.526
paD2A	0.02	0.0010	2.88×10^{-2}	1.000	0.900	0.500
taBEE	0.03	0.1286	1.04×10^{4}	0.914	0.904	0.486
paBEE	0.02	0.0439	1.31×10^{-1}	1.000	0.893	0.497

Table 5. *Cont.*

Parameters	R^2	Bias	RMSE	Factor 2	Coverage 90%	Coverage 50%
taD1P	0.11	−0.0070	1.72×10^4	0.973	0.897	0.499
paD1P	0.02	−0.0002	2.85×10^{-2}	1.000	0.897	0.508
taARP	0.15	−0.0002	1.85×10^4	0.988	0.916	0.517
paARP	0.03	−0.0014	2.85×10^{-2}	1.000	0.906	0.509
taNEA	0.10	−0.0204	1.06×10^4	0.992	0.893	0.516
paNEA	0.02	0.0000	2.81×10^{-2}	1.000	0.924	0.516
taNG2	0.15	−0.0223	1.36×10^4	0.998	0.909	0.528
paNG2	0.02	−0.0003	2.89×10^{-2}	1.000	0.909	0.477
mYG1	0.15	1.2696	2.69×10^{-4}	0.709	0.927	0.521
mG1Y	0.03	1.8171	2.86×10^{-4}	0.742	0.907	0.516
mG1G2	0.05	2.0667	2.85×10^{-4}	0.737	0.895	0.519
mG2G1	0.05	2.9954	2.89×10^{-4}	0.745	0.885	0.509
mG2E	0.03	3.0547	3.01×10^{-4}	0.692	0.886	0.460
mEG2	0.19	1.5013	2.67×10^{-4}	0.722	0.908	0.503
mEA	0.12	1.4834	2.68×10^{-4}	0.744	0.902	0.543
mAE	0.11	1.9813	2.74×10^{-4}	0.731	0.908	0.523
mAP	0.27	1.4789	2.40×10^{-4}	0.766	0.910	0.548
mPA	0.37	2.2687	2.35×10^{-4}	0.773	0.908	0.546
m1G2EA	0.02	2.1201	2.90×10^{-4}	0.701	0.911	0.489
m1EAG2	0.04	2.7879	2.92×10^{-4}	0.708	0.888	0.496
m1EAP	0.06	2.5111	2.82×10^{-4}	0.728	0.901	0.528
m1PEA	0.05	3.2113	2.91×10^{-4}	0.694	0.911	0.477

The estimates for the current African effective population size (nY) is about 15,000 (median value), in agreement with previous studies [37,38]. A lower value is estimated for the Eurasians, with an effective population size of about 7000 individuals for the Europeans (nE) and of about 11,000 individuals for the Asians (nA). A bit higher is the estimate for Australo-Melanesian population: the median value of the effective population size is indeed about 25,000 individuals (nP).

The first divergence within Africa (tdYG1), that generated the source population giving rise to the first wave of migrants has been estimated about 104,000 years ago, with a 95% confidence interval between 55,000 and 141,000 years ago (and a 50% CI between 78,000 and 125,000 years ago). The first waves of migrants left Africa (tdOA1) about 74,000 years ago (95% CI: 47,000–120,000 years ago; 50% CI: 55,000–96,000 years ago), whereas the second wave of migration (tdOA2), originated from a structure generated (tdYG2) about 100,000 years ago, left Africa about 46,000 years ago (95% CI: 40,000–59,000 years ago, 50% CI: 42,000–51,000 years ago). Europeans and Asians diverged (tdEA) about 37,000 years ago. These estimates are in agreement with a previous work that considered a less realistic model and a smaller amount of genetic data [11].

4. Discussion

In this paper, we explicitly compared two models of AMH evolution through an ABC–RF approach based on the analysis of modern and ancient complete genomes. The two tested demographic models consider details of our evolutionary history that have been proposed in the recent literature, such as the presence of a (so far, unsampled) Basal European population contributing to the genome of recent Europeans [30], or the two distinct pulses of admixture from two different Denisovan populations to Asians and Papuans [23]. The main difference between the two scenarios regards the dynamics of expansion from Africa of AMH. According to the SD model, all non-African populations derive from a single major migration wave; on the contrary, the MD model assumes two migration waves, distinct in time and place, the first one giving rise to modern Australo-Melanesians and the other giving rise to Eurasians. Needless to say, successive processes of gene flow and admixture have certainly complicated the apparently simple patterns generated by the initial African dispersal(s). Yet, even these admittedly simplified models are complex (defined by up to 50 parameters), and the differences between them are relatively small; therefore, one could expect that it might be difficult to tell them apart. On the contrary, the ABC-RF procedure we chose provided a good discriminatory power, with a proportion of True Positives of about 70% for both AD and MD models. This TP proportion is comparable to, or higher than, that reported in previous works where simpler (and hence less realistic) models were analyzed (see e.g., [39,40]). When the two alternative models were compared, the MD model resulted consistently four-fold more probable than the SD model, no matter which Papuan (Table 2), African, European or Asian individuals were considered (Table S4), with a posterior probability estimated around 80%. The support for the MD model is marginally higher than in [16], where a comparison between two alternative, less up-to-date, evolutionary histories of AMH favored the MD model with a probability of about 75%. These results are robust to slight changes in the MD parametrization. We indeed tested also a version of MD in which Papuans derived part of their genomes from Eurasians, modeled as a single pulse of admixture occurring after the second exit (rather than through a process of continuous gene flow), the results are reported in Table S7. Even in this version, the MD appeared more supported by data than the SD model, although it appeared slightly less likely than the previous MD model when included in the general comparison.

In this work, for the first time, we also attempted to estimate the parameters of the supported model by ABC-RF. The MD model was defined by 50 free parameters, estimated through the regression random forest algorithm [20]. We also assessed the quality of these estimates through the calculation of statistics that gave us information about the inferential power of the parameter's estimation procedure. An assessment of the quality of the estimated parameters was prohibitive so far, due to computational limits of other inferential methods, e.g., those based on composite-likelihood [41]. With ABC-RF, instead, the same reference table (made up of just a few thousand simulations) allows one to both estimate parameters and assess their quality using a subset of the simulation as "pods". To perform the same analysis by composite-likelihood methods, one would require about 100 thousand new simulations for each pod analyzed, which means, even considering only 100 pods, billions of simulations. This large amount of simulated data often exceeds computational constraints, in particular when complex demographies are analyzed. As a consequence, in studies of complex models, no information was provided about the reliability of parameter estimates [13,42]. The procedure we applied made it possible to compensate for this drawback, as shown in Table 5.

It would have been unrealistic to expect that all 50 parameters could be reliably estimated. The migration rates among modern populations, or the proportion and timing of admixture events, for instance, proved elusive, showing a low R^2 and high bias and RMSE values. We knew that there is an almost infinite set of parameter combinations leading to the same patterns of genome diversity, with, for instance, old small-scale admixture events, and recent larger-scale admixture events, producing, in principle, the same consequences at the genomic level. Other parameters show better estimates. This is the case of the effective population sizes, or, to a lesser extent, of the divergence times. The African, European and Asian estimates of the effective population sizes are consistent with what reported in the literature [38,43]; the higher value estimated for the Australo-Melanesian group, here represented by the Papuans, may be surprising, but it is in agreement with the harmonic mean of the effective population sizes estimated over time by [12].

The most interesting parameters are those associated with the divergence/departure from Africa. These parameters show R^2 above 10%, good coverage, and a factor 2 of about 100%; however, their confidence intervals are huge and their posterior distributions often seem to reflect the prior range. This means that we should still take with caution these estimates and that the ABC inferential procedure, albeit powerful, shows room for improvement. The key advantage of the ABC estimation is that the "quality assessment" procedure allows the acquisition of consciousness about the quality of the estimates; nevertheless, having this in mind, we can still discuss the estimates obtained. We dated the structure of African groups that gave rise to the source populations of the migration waves from Africa about 100,000 years ago. The bottleneck of the first exit from Africa, associated with the origin of Australo-Melanesian groups, has been estimated at about 74,000 years ago, in line with the timing inferred from paleoanthropological data (70,000 years ago, [44]). The second exit, giving rise to Eurasian populations, was placed at about 46,000 years ago. This is in agreement with previous estimates from genomic data [4,38,45] and receives further support from the relatively recent arrival of modern humans in Europe suggested by much of the archaeological evidence (40–45 thousand years ago, [46,47]). Some authors proposed an even earlier presence of AMH in Europe [48]. Be that as it may, it is also plausible that large-scale gene flow processes, documented at least twice in Europe (in the Neolithic period and Bronze Age; see [49]) may have slightly reduced diversity and hence the apparent depth of the DNA genealogies, thus producing a bias towards more recent values in the estimation of divergence times. The two migration waves from Africa considered in the MD model appear to be separated in time, with no temporal overlap considering their 50% confidence interval (55,000–96,000 for the first exit and 42,000–51,000 for the second exit), and a limited overlap considering their 95% confidence interval (47,000–120,000 for the first exit and 40,000–59,000 for the second exit).

5. Conclusions

In this paper we extensively tested two up-to-date models of modern human expansion Out of Africa through a machine learning ABC approach. The simulated variation has been compared with those observed in ancient and modern genomes, and our results consistently supported a Multiple Dispersal Model, in which modern Australo-Melanesians derive from an earlier migration from Africa than that giving rise to Eurasians. We also estimated the parameters of the most supported model, and we concentrated our effort in assessing the quality of the estimates produced. This procedure, albeit fundamental to ensure the reliability of the estimates, it is rarely performed, due to the limitations of available inferential methods. These limitations are currently overcame by the ABC-RF procedure coupled with the FDSS statistic, which allowed us to highlight weakness and strengths of the parameters estimated. Our results indeed support that the hypothesis of two main dispersal event from Africa, separated in time and place [10–12], cannot be dismissed [4,13], but the quality assessment of the parameters we estimated certainly show that needs to be further explored.

Supplementary Materials: The following are available online at http://www.mdpi.com/2073-4425/11/12/1510/s1, Table S1: Demographic parameters and prior distributions of Single Dispersal model. Table S2: Demographic parameters and prior distributions of Multiple Dispersal model. Table S3: Complete list of genomes used for the comparison of Single Dispersal model and Multiple Dispersal model using real data; Table S4: Results of model selection performed using alternative individuals from African, European and Asian populations; Table S5: Power test of model comparison for increasing number of simulations considered in the reference table.; Table S6. Complete list of acronyms of the MD model's demographic parameters.; Table S7. Model Selection results including the MD-Pulse admixture model. Figure S1: Outline of the entire workflow; Figure S2: Posterior density of the effective population sizes estimated using the Papuan sample from Malaspinas et al. (2016). Figure S3: Posterior density of the divergence times and the admixture times estimated using the Papuan sample from Malaspinas et al. (2016). Figure S4: Posterior density of the admixture rates estimated using the Papuan sample from Malaspinas et al. (2016). Figure S5: Posterior density of the migration rates estimated using the Papuan sample from Malaspinas et al. (2016). Figure S6: Posterior density of the effective population sizes estimated using the Papuan sample from Pagani et al. (2016). Figure S7: Posterior density of the divergence times and the admixture times estimated using the Papuan sample from Pagani et al. (2016). Figure S8: Posterior density of the admixture rates estimated using the Papuan sample from Pagani et al. (2016). Figure S9: Posterior density of the migration rates estimated using the Papuan sample from Pagani et al. (2016). Figure S10: The model below represents a simplified version of the most supported model (MD) showing the main demographic parameters.

Author Contributions: Conceptualization, G.B. and S.G.; formal analysis, M.T.V.; methodology, A.B. and S.G.; software, M.T.V. and A.B.; supervision, G.B. and S.G.; writing—original draft, G.B. and S.G.; writing—review and editing, M.T.V., A.B., G.B., and S.G. All authors have read and agreed to the published version of the manuscript.

Funding: This research received no external funding.

Acknowledgments: We are indebted to Francesca Tassi and Alberto Seno for technical help.

Conflicts of Interest: The authors declare no conflict of interest.

References

1. Scerri, E.M.L.; Thomas, M.G.; Manica, A.; Gunz, P.; Stock, J.T.; Stringer, C.; Grove, M.; Groucutt, H.S.; Timmermann, A.; Rightmire, G.P.; et al. Did Our Species Evolve in Subdivided Populations across Africa, and Why Does It Matter? *Trends Ecol. Evol.* **2018**, *33*, 582–594. [CrossRef]
2. Mellars, P. Neanderthals and the Modern Human Colonization of Europe. *Nature* **2004**, *432*, 461–465. [CrossRef]
3. Higham, T.; Douka, K.; Wood, R.; Ramsey, C.B.; Brock, F.; Basell, L.; Camps, M.; Arrizabalaga, A.; Baena, J.; Barroso-Ruíz, C.; et al. The Timing and Spatiotemporal Patterning of Neanderthal Disappearance. *Nature* **2014**, *512*, 306–309. [CrossRef] [PubMed]
4. Mallick, S.; Li, H.; Lipson, M.; Mathieson, I.; Gymrek, M.; Racimo, F.; Zhao, M.; Chennagiri, N.; Nordenfelt, S.; Tandon, A.; et al. The Simons Genome Diversity Project: 300 Genomes from 142 Diverse Populations. *Nature* **2016**, *538*, 201–206. [CrossRef] [PubMed]
5. Hershkovitz, I.; Weber, G.W.; Quam, R.; Duval, M.; Grün, R.; Kinsley, L.; Ayalon, A.; Bar-Matthews, M.; Valladas, H.; Mercier, N.; et al. The Earliest Modern Humans Outside Africa. *Science* **2018**, *359*, 456–459. [CrossRef]
6. Liu, H.; Prugnolle, F.; Manica, A.; Balloux, F. A Geographically Explicit Genetic Model of Worldwide Human-Settlement History. *Am. J. Hum. Genet.* **2006**, *79*, 230–237. [CrossRef] [PubMed]
7. Mellars, P.; Gori, K.C.; Carr, M.; Soares, P.A.; Richards, M.B. Genetic and Archaeological Perspectives on the Initial Modern Human Colonization of Southern Asia. *Proc. Natl. Acad. Sci. USA* **2013**, *110*, 10699–10704. [CrossRef] [PubMed]
8. López, S.; Van Dorp, L.; Hellenthal, G. Human Dispersal out of Africa: A Lasting Debate. *Evol. Bioinform.* **2015**. [CrossRef] [PubMed]
9. Lahr, M.M.; Foley, R. Multiple Dispersals and Modern Human Origins. *Evol. Anthropol. Issues News Rev.* **1994**, *3*, 48–60. [CrossRef]
10. Reyes-Centeno, H.; Ghirotto, S.; Detroit, F.; Grimaud-Herve, D.; Barbujani, G.; Harvati, K. Genomic and Cranial Phenotype Data Support Multiple Modern Human Dispersals from Africa and a Southern Route into Asia. *Proc. Natl. Acad. Sci. USA* **2014**, *111*, 7248–7253. [CrossRef]
11. Tassi, F.; Ghirotto, S.; Mezzavilla, M.; Vilaça, S.T.; De Santi, L.; Barbujani, G. Early Modern Human Dispersal from Africa: Genomic Evidence for Multiple Waves of Migration. *Investig. Genet.* **2015**, *6*, 6–13. [CrossRef] [PubMed]
12. Pagani, L.; Lawson, D.J.; Jagoda, E.; Mörseburg, A.; Eriksson, A.; Mitt, M.; Clemente, F.; Hudjashov, G.; DeGiorgio, M.; Saag, L.; et al. Genomic Analyses Inform on Migration Events during the Peopling of Eurasia. *Nature* **2016**, *538*, 238–242. [CrossRef] [PubMed]
13. Malaspinas, A.S.; Westaway, M.C.; Muller, C.; Sousa, V.C.; Lao, O.; Alves, I.; Bergström, A.; Georgios, A.; Cheng, J.Y.; Crawford, G.E. A Genomic History of Aboriginal Australia. *Nature* **2016**, *538*, 207–214. [CrossRef] [PubMed]
14. Varin, C. On Composite Marginal Likelihoods. *Asta Adv. Stat. Anal.* **2008**, *92*, 1–28. [CrossRef]
15. Varin, C.; Reid, N.; Firth, D. An Overview of Composite Likelihood Methods. *Stat. Sin.* **2011**, *21*, 5–42.
16. Ghirotto, S.; Vizzari, M.T.; Tassi, F.; Barbujani, G.; Benazzo, A. Distinguishing among Complex Evolutionary Models Using Unphased Whole-genome Data through Random-Forest Approximate Bayesian Computation. *Mol. Ecol. Resour.* **2020**, 1–15. [CrossRef]
17. Beaumont, M.A.; Zhang, W.; Balding, D.J. Approximate Bayesian Computation in Population Genetics. *Genetics* **2002**, *162*, 2025–2035.
18. Beaumont, M.A. Joint Determination of Topology, Divergence Time, and Immigration in Population Trees. In *Simulations, Genetics and Human Prehistory*; McDonald Institute for Archaeological Research: Cambridge, UK, 2008; pp. 135–154.
19. Pudlo, P.; Marin, J.M.; Estoup, A.; Cornuet, J.M.; Gautier, M.; Robert, C.P. Reliable ABC Model Choice via Random Forests. *Bioinformatics* **2015**, *32*, 859–866. [CrossRef] [PubMed]
20. Raynal, L.; Marin, J.M.; Pudlo, P.; Ribatet, M.; Robert, C.P.; Estoup, A. ABC Random Forests for Bayesian Parameter Inference. *Bioinformatics* **2019**, *35*, 1720–1728. [CrossRef]
21. Wakeley, J.; Hey, J. Estimating Ancestral Population Parameters. *Genetics* **1997**, *145*, 847–855.

22. Mondal, M.; Casals, F.; Xu, T.; Dall'Olio, G.M.; Pybus, M.; Netea, M.G.; Comas, D.; Laayouni, H.; Li, Q.; Majumder, P.P.; et al. Genomic Analysis of Andamanese Provides Insights into Ancient Human Migration into Asia and Adaptation. *Nat. Genet.* **2016**, *48*, 1066–1070. [CrossRef] [PubMed]
23. Browning, S.R.; Browning, B.L.; Zhou, Y.; Tucci, S.; Akey, J.M. Analysis of Human Sequence Data Reveals Two Pulses of Archaic Denisovan Admixture. *Cell* **2018**, *173*, 53–61.e9. [CrossRef]
24. Jacobs, G.S.; Hudjashov, G.; Saag, L.; Kusuma, P.; Darusallam, C.C.; Lawson, D.J.; Mondal, M.; Pagani, L.; Ricaut, F.-X.; Stoneking, M.; et al. Multiple Deeply Divergent Denisovan Ancestries in Papuans. *Cell* **2019**, *177*, 1010–1021. [CrossRef] [PubMed]
25. Wall, J.D.; Yang, M.A.; Jay, F.; Kim, S.K.; Durand, E.Y.; Stevison, L.S.; Gignoux, C.; Woerner, A.; Hammer, M.F.; Slatkin, M. Higher Levels of Neanderthal Ancestry in East Asians than in Europeans. *Genetics* **2013**, *194*, 199–209. [CrossRef] [PubMed]
26. Prüfer, K.; Racimo, F.; Patterson, N.; Jay, F.; Sankararaman, S.; Sawyer, S.; Heinze, A.; Renaud, G.; Sudmant, P.H.; De Filippo, C.; et al. The Complete Genome Sequence of a Neanderthal from the Altai Mountains. *Nature* **2014**, *505*, 43–49. [CrossRef] [PubMed]
27. Vernot, B.; Akey, J.M. Resurrecting Surviving Neandertal Lineages from Modern Human Genomes. *Science* **2014**, *343*, 1017–1021. [CrossRef] [PubMed]
28. Lazaridis, I.; Patterson, N.; Mittnik, A.; Renaud, G.; Mallick, S.; Kirsanow, K.; Sudmant, P.H.; Schraiber, J.G.; Castellano, S.; Lipson, M.; et al. Ancient Human Genomes Suggest Three Ancestral Populations for Present-Day Europeans. *Nature* **2014**, *513*, 409–413. [CrossRef]
29. Lazaridis, I.; Nadel, D.; Rollefson, G.; Merrett, D.C.; Rohland, N.; Mallick, S.; Fernandes, D.; Novak, M.; Gamarra, B.; Sirak, K.; et al. Genomic Insights into the Origin of Farming in the Ancient Near East. *Nature* **2016**, *536*, 419–424. [CrossRef]
30. Villanea, F.A.; Schraiber, J.G. Multiple Episodes of Interbreeding between Neanderthal and Modern Humans. *Nat. Ecol. Evol.* **2019**, *3*, 39–44. [CrossRef]
31. Scally, A.; Durbin, R. Revising the Human Mutation Rate: Implications for Understanding Human Evolution. *Nat. Rev. Genet.* **2012**, *13*, 745–753. [CrossRef]
32. Hudson, R.R. Generating Samples under a Wright-Fisher Neutral Model of Genetic Variation. *Bioinformatics* **2002**, *18*, 337–338. [CrossRef]
33. Meyer, M.; Kircher, M.; Gansauge, M.T.; Li, H.; Racimo, F.; Mallick, S.; Schraiber, J.G.; Jay, F.; Prüfer, K.; De Filippo, C.; et al. A High-Coverage Genome Sequence from an Archaic Denisovan Individual. *Science* **2012**, *338*, 222–226. [CrossRef]
34. Hinrichs, A.S.; Raney, B.J.; Speir, M.L.; Rhead, B.; Casper, J.; Karolchik, D.; Kuhn, R.M.; Rosenbloom, K.R.; Zweig, A.S.; Haussler, D.; et al. UCSC Data Integrator and Variant Annotation Integrator. *Bioinformatics* **2016**, *32*, 1430–1432. [CrossRef]
35. Li, H.; Handsaker, B.; Wysoker, A.; Fennell, T.; Ruan, J.; Homer, N.; Marth, G.; Abecasis, G.; Durbin, R. The Sequence Alignment/Map Format and SAMtools. *Bioinformatics* **2009**, *25*, 2078–2079. [CrossRef] [PubMed]
36. Neuenschwander, S.; Largiadèr, C.R.; Ray, N.; Currat, M.; Vonlanthen, P.; Excoffier, L. Colonization History of the Swiss Rhine Basin by the Bullhead (Cottus Gobio): Inference under a Bayesian Spatially Explicit Framework. *Mol. Ecol.* **2008**, *17*, 757–772. [CrossRef] [PubMed]
37. Fan, S.; Kelly, D.E.; Beltrame, M.H.; Hansen, M.E.B.; Mallick, S.; Ranciaro, A.; Hirbo, J.; Thompson, S.; Beggs, W.; Nyambo, T.; et al. African Evolutionary History Inferred from Whole Genome Sequence Data of 44 Indigenous African Populations. *Genome Biol.* **2019**, *20*, 1–14.
38. McEvoy, B.P.; Powell, J.E.; Goddard, M.E.; Visscher, P.M. Human Population Dispersal "Out of Africa" Estimated from Linkage Disequilibrium and Allele Frequencies of SNPs. *Genome Res.* **2011**, *21*, 821–829. [CrossRef]
39. Fagundes, N.J.R.; Ray, N.; Beaumont, M.; Neuenschwander, S.; Salzano, F.M.; Bonatto, S.L.; Excoffier, L. Statistical Evaluation of Alternative Models of Human Evolution. *Proc. Natl. Acad. Sci. USA* **2007**, *104*, 17614–17619. [CrossRef] [PubMed]
40. Veeramah, K.R.; Wegmann, D.; Woerner, A.; Mendez, F.L.; Watkins, J.C.; Destro-Bisol, G.; Soodyall, H.; Louie, L.; Hammer, M.F. An Early Divergence of KhoeSan Ancestors from Those of Other Modern Humans Is Supported by an ABC-Based Analysis of Autosomal Resequencing Data. *Mol. Biol. Evol.* **2012**, *29*, 617–630. [CrossRef]

41. Excoffier, L.; Dupanloup, I.; Huerta-Sánchez, E.; Sousa, V.C.; Foll, M. Robust Demographic Inference from Genomic and SNP Data. *PLoS Genet.* **2013**, *9*, e1003905. [CrossRef]
42. Nater, A.; Mattle-Greminger, M.P.; Nurcahyo, A.; Nowak, M.G.; De Manuel, M.; Desai, T.; Groves, C.; Pybus, M.; Sonay, T.B.; Roos, C.; et al. Morphometric, Behavioral, and Genomic Evidence for a New Orangutan Species. *Curr. Biol.* **2017**, *27*, 3576–3577. [CrossRef] [PubMed]
43. Schiffels, S.; Durbin, R. Inferring Human Population Size and Separation History from Multiple Genome Sequences. *Nat. Genet.* **2014**, *46*, 919–925. [CrossRef] [PubMed]
44. Mirazón Lahr, M.; Foley, R.A. Towards a Theory of Modern Human Origins: Geography, Demography, and Diversity in Recent Human Evolution. *Am. J. Phys. Anthropol.* **1999**, *107*, 137–176. [CrossRef]
45. Gravel, S.; Henn, B.M.; Gutenkunst, R.N.; Indap, A.R.; Marth, G.T.; Clark, A.G.; Yu, F.; Gibbs, R.A.; Bustamante, C.D.; The 1000 Genomes Project; et al. Demographic History and Rare Allele Sharing among Human Populations. *Proc. Natl. Acad. Sci. USA* **2011**, *108*, 11983–11988. [CrossRef] [PubMed]
46. Mellars, P. Why Did Modern Human Populations Disperse from Africa ca. 60,000 Years Ago? A New Model. *Proc. Natl. Acad. Sci. USA* **2006**, *103*, 9381–9386. [CrossRef]
47. Reyes-Centeno, H.; Hubbe, M.; Hanihara, T.; Stringer, C.; Harvati, K. Testing Modern Human Out-of Africa Dispersal Models and Implications for Modern Human Origins. *J. Hum. Evol.* **2015**, *87*, 95–106. [CrossRef]
48. Hublin, J.J.; Sirakov, N.; Aldeias, V.; Bailey, S.; Bard, E.; Delvigne, V.; Endarova, E.; Fagault, Y.; Fewlass, H.; Hajdinjak, M.; et al. Initial Upper Palaeolithic Homo Sapiens from Bacho Kiro Cave, Bulgaria. *Nature* **2020**, *581*, 299–302. [CrossRef]
49. Haak, W.; Lazaridis, I.; Patterson, N.; Rohland, N.; Mallick, S.; Llamas, B.; Brandt, G.; Nordenfelt, S.; Harney, E.; Stewardson, K.; et al. Massive Migration from the Steppe Was a Source for Indo-European Languages in Europe. *Nature* **2015**, *522*, 207–211. [CrossRef]

MDPI
St. Alban-Anlage 66
4052 Basel
Switzerland
Tel. +41 61 683 77 34
Fax +41 61 302 89 18
www.mdpi.com

Genes Editorial Office
E-mail: genes@mdpi.com
www.mdpi.com/journal/genes

www.ingramcontent.com/pod-product-compliance
Lightning Source LLC
LaVergne TN
LVHW070625170726
843515LV00004B/181

* 9 7 8 3 0 3 6 5 2 5 4 7 1 *